Sir Charles Bell

Advance Praise for *Sir Charles Bell*

"With warmth and knowledge, Michael Aminoff provides a well-rounded summary of the life and career of Sir Charles Bell. The reader enters the world of early 19th century London, travels alongside Bell through the scientific and medical avenues of descriptive neuroanatomy, and enters several darker alleys of controversy. The journey unveils provocative historical questions and new insights related to this important, largely-forgotten scientific and medical figure."
—*Christopher G. Goetz, MD, Professor of Neurology, Rush University Medical Center, Chicago, IL*

"Had I been asked who Sir Charles Bell was, I would have simply replied, 'The guy who described Bell's palsy.' Now, after reading Michael Aminoff's book, I realize that Bell's palsy was a very minor aspect of Bell's extraordinary accomplishments. He was a surgeon, neurologist, anatomist, prolific writer, and illustrator. Indeed, he was the ultimate polymath. Bell's book *Idea of a New Anatomy of the Brain* was described as the Magna Carta of neurology. Dr. Aminoff, also a polymath, captures Bell's true genius in a book that will be enjoyed by all seekers of true brilliance."
—*Robert B. Daroff, MD, Professor and Chair Emeritus of Neurology,*
Case Western Reserve University School of Medicine, Cleveland, OH

"From a forgotten corner – modern neurology's birth in the early 19th century – Michael Aminoff restores Charles Bell to our consciousness as a founding father. This is vibrant reading, providing a rich view of this brilliant, driven, and flawed man and his moment in time. Aminoff offers a balanced view of Bell's foibles as well as his prodigious accomplishments. He deftly shows us how so many of the clinical syndromes, mapped in modern form over the next century, were actually recognized by Bell and his contemporaries. Kudos to Michael Aminoff for bringing Bell out of the shadows of history."
—*Stephen L. Hauser, MD, Director, UCSF Weill Institute for Neurosciences, Professor & Chair,*
Department of Neurology, University of California San Francisco, San Francisco, CA

"Michael Aminoff is a renowned educator of neurology and neurological sciences. His writing is eclectic and this latest monograph tackles the life, works, and thoughts of Sir Charles Bell. It is by far the most detailed description of Bell ever written, full of fascinating details. Aminoff is a physician-scientist of considerable stature and this had given him the ability to weave Bell's discoveries into present day thinking and to indicate how observations made by Bell have significant relevance in modern medicine. This monograph delves into England and Europe as experienced by Bell, adding charm and interest for non-physicians. Aminoff is not shy of bringing out Charles Bell's shortcomings. Despite this I am sure that Sir Charles would thoroughly enjoy this gem, full of wonderful descriptions of his many achievements."
—*Andrew Eisen, MD, FRCPC, Professor Emeritus, Division of Neurology,*
University of British Columbia, Vancouver, Canada

"This volume will be a pleasure for all of us in search of an important scientific biography that reads like a novel. A reference text in its own right, it is an intimate account of Charles Bell through his life, work, and art, and also a rich cultural history of the times in which he lived. Through extensive research, Michael Aminoff finds the humanity in a brilliant and important, but flawed figure at the birth of neuroscience."
—*Kerry H. Levin, MD, Chairman, Department of Neurology, Cleveland Clinic, Cleveland, OH*

"It is difficult to imagine many clinical neurologists who have not heard of Sir Charles Bell or at least of some of the clinical phenomena and neurological disorders that bear his name. In this book, Michael Aminoff makes a compelling case for the importance of studying those who preceded us. He gives insights into a fascinating and prolific man, as well as the backdrop of his times. The crude nature of medical knowledge and treatments at that time highlights just how remarkable were Bell's insights into the structure and organization of the nervous system. This is a nice piece of work … a very readable book."
—*A. Jon Stoessl, MD, FRCPC, Professor & Head of Neurology,*
University of British Columbia, Vancouver, Canada

Sir Charles Bell

His Life, Art, Neurological Concepts, and Controversial Legacy

Michael J. Aminoff, MD, DSc, FRCP

Professor of Neurology
School of Medicine
University of California
San Francisco, California

UNIVERSITY PRESS

Oxford University Press is a department of the University of Oxford. It furthers
the University's objective of excellence in research, scholarship, and education
by publishing worldwide. Oxford is a registered trade mark of Oxford University
Press in the UK and certain other countries.

Published in the United States of America by Oxford University Press
198 Madison Avenue, New York, NY 10016, United States of America.

© Michael J. Aminoff 2017

All rights reserved. No part of this publication may be reproduced, stored in
a retrieval system, or transmitted, in any form or by any means, without the
prior permission in writing of Oxford University Press, or as expressly permitted
by law, by license, or under terms agreed with the appropriate reproduction
rights organization. Inquiries concerning reproduction outside the scope of the
above should be sent to the Rights Department, Oxford University Press, at the
address above.

You must not circulate this work in any other form
and you must impose this same condition on any acquirer.

Library of Congress Cataloging-in-Publication Data
Names: Aminoff, Michael J. (Michael Jeffrey), author.
Title: Sir Charles Bell : his life, art, neurological concepts, and controversial legacy /
by Michael J. Aminoff.
Description: Oxford ; New York : Oxford University Press, [2017] |
Includes bibliographical references and index.
Identifiers: LCCN 2016013519 | ISBN 9780190614966 (paper)
Subjects: | MESH: Bell, Charles, Sir, 1774–1842. | Surgeons | Surgeons—history | Anatomy—history |
Neurophysiology—history | History, 19th Century | England | Biography
Classification: LCC RD27.35.B435 | NLM WZ 100 | DDC 617.092—dc23
LC record available at http://lccn.loc.gov/2016013519

This material is not intended to be, and should not be considered, a substitute for medical or other
professional advice. Treatment for the conditions described in this material is highly dependent on
the individual circumstances. And, while this material is designed to offer accurate information with
respect to the subject matter covered and to be current as of the time it was written, research and
knowledge about medical and health issues is constantly evolving and dose schedules for medications
are being revised continually, with new side effects recognized and accounted for regularly. Readers
must therefore always check the product information and clinical procedures with the most up-to-date
published product information and data sheets provided by the manufacturers and the most recent
codes of conduct and safety regulation. The publisher and the authors make no representations or
warranties to readers, express or implied, as to the accuracy or completeness of this material. Without
limiting the foregoing, the publisher and the authors make no representations or warranties as to the
accuracy or efficacy of the drug dosages mentioned in the material. The authors and the publisher do
not accept, and expressly disclaim, any responsibility for any liability, loss or risk that may be claimed or
incurred as a consequence of the use and/or application of any of the contents of this material.

To Jan, my wife for more than forty years, with love, respect, and gratitude

CONTENTS

Preface ix
Acknowledgments xiii

1. An Introductory Snapshot 1
2. In the Beginning 5
3. London and the Great Windmill Street School 19
4. Anatomy of the Expression of Emotions 31
5. Behind the Glories of War 57
6. Swings and Roundabouts 77
7. In and Out of the Central Nervous System 97
8. The Organization of the Nervous System 117
9. Clinical Observations on the Nervous System 131
10. For God and Country 145
11. New Classrooms: Old Struggles 165
12. The Ebbing Tide 189
13. To Each His Due 201

Appendix 1. Books Authored by Charles Bell 209
Appendix 2. Charles Bell's Published Papers and Lectures 219
Appendix 3. The Drawings and Paintings of Charles Bell 223
Appendix 4. *Idea of a New Anatomy of the Brain* 227
About the Author 237
Index 239

PREFACE

In the early nineteenth century, a relatively obscure general surgeon wrote a small, privately published booklet that was so thoughtful and thought-provoking that it has been described by some as the *Magna Carta* of neurology. Why, then, is it unknown today by most neuroscientists, biologists, and clinical neurologists? It is curious that its author, a man whose achievements have come to be likened to those of William Harvey, is also all but forgotten. Does he deserve the indifference of the medical and scientific community, a community that actually owes much to his foresight? Is he forgotten in the same way as so many others whose contributions have led to an expansion of human knowledge, simply because fame is ephemeral? Or is he ignored because—in retrospect—his conduct has been judged by some to have breached the standard of intellectual honesty that is expected of those to whom we look for example? It was questions of this sort that prompted me to study the life and work of Charles Bell. In doing so, I was struck by the multitude of his accomplishments as a clinician, scientist, artist, and educator, and I came to believe that these accomplishments—and his failings—deserve to be better known. This, in turn, prompted me to write the present appraisal of Bell and his work. Present-day scientists and educators, surgeons and artists, would do well to appreciate his achievements and thereby gain a better understanding of the foundations on which they themselves build as well as a greater awareness of the place of an individual in the development of biological concepts and ideas. At the same time, an understanding of the fragility beneath his mask of polished professionalism gives Bell a human dimension—warts and all—to which everyone can relate.

It is perhaps surprising that Bell's life and work has not been studied in more detail. A number of brief essays and articles have appeared in scholarly journals, but these generally focus on only one aspect or another of his work, are often based on secondary sources, and offer limited insight into his achievements and failings. The full-length biography of him by Gordon-Taylor and Walls that was published in 1958 is now long out of print. Just before this book was completed, I became aware of a new book by Carin Berkowitz, then in press and subsequently published, titled *Charles Bell and the Anatomy of Reform*. However, it focuses especially on the London medical classroom and on Bell's use of educational materials and institutions to disseminate scientific knowledge. The present volume is intended to provide a broader account of the life and work of Charles Bell that will—I hope—be of interest to physicians, surgeons, anatomists, physiologists, artists, psychologists, educators, and medical historians, as well as to a certain segment of the lay public. I have been concerned to present not only his achievements and successes but also the less flattering aspects of his character and professional conduct, including his single-minded push to further his own reputation even at the expense of others. It is my hope that I have been able to do so while—at the same time—placing his ideas in their cultural and historical contexts.

Some, especially scientists, may question the necessity of looking to the past in an age of rapid advances in scope and depth of all branches of human knowledge. Indeed, the essayist Clive James, in a memoir on Charles Chaplin, commented that science "lives in a perpetual present," discarding its own past as it advances, in contrast to the humanities, which accumulate and always retain the past. An understanding of past concepts and endeavors provides an informed framework to modern studies, however, allowing a better perspective on contemporary issues. Advances or changes—whether in the arts or sciences—also gain more meaning when they can be associated with a human face and are understood in the context of the times in which they occurred. To neglect the legacy of thought left by past generations is to squander a potential treasure trove of ideas that may be as relevant today as when they were originally formulated. Sir Andrew Huxley, the Nobel laureate, in his 1982 Florey Lecture (published in *Proceedings of the Royal Society of London, Series B*, Vol. 216, pp. 253–266) recalled that some chance experimental findings of his might have been obtained much earlier by planned experiments if he had been familiar with the literature of the nineteenth century. Indeed, he wondered how many more important suggestions "are still sitting unknown in the massive literature" of that time.

Given the many facets of Bell's professional life, it has been difficult to follow a strictly chronological course in this account without confusing the reader, so the book is also arranged thematically. Although I have taken the main events chronologically, I have then followed them through to the end so that each topic is discussed coherently and comprehensively in one place. For example, in discussing his work in relation to the arts, Chapter 4 provides an account of his influential book on the anatomy of expression, first published in 1806, and its consequences over the succeeding years up to the present, whereas Chapter 5 discusses his famous paintings and drawings of the wounded in the Napoleonic Wars in 1808 but also in 1815 (Waterloo). General or anecdotal information about important contemporaries, institutions, concepts of disease, and beliefs is provided in an easily accessible format in footnotes to the text rather than at the end of the chapter or book. In addition, an extensive bibliography is provided separately at the end of each chapter to allow the reader to pursue topics of particular interest. I have endeavored to place Bell's contributions in the context of the times but have also discussed subsequent developments in the field to show their relevance at the present time. I hope that this will allow his contributions to receive the consideration that they deserve.

On a number of occasions, I have quoted from Bell's published correspondence with his brother. I must emphasize, therefore, that this correspondence was published by his widow almost thirty years after his death, and the extent to which his letters have been expurgated or edited is unclear. Furthermore, in preparing the present volume, I have made no attempt to express the value of nineteenth-century goods, services, or incomes in present-day values because any computation depends on which of several different measures (such as the consumer price index, gross domestic product, or unskilled wage rate) is used.

I was a medical student at University College London—the original University of London that Bell helped to found and where he was the most famous of the faculty originally appointed to the medical department—in the late 1950s and early 1960s. I vividly recall spending hours in the dissecting room working over an embalmed human body, and also remember the students from the Slade School of Fine Arts joining us for anatomy classes, just as Bell had once urged. I also worked for almost two years as a registrar in neurology at the Middlesex Hospital, where Bell—more than one hundred and fifty years earlier—had been on the surgical staff for some years. As a neurologist and clinical neurophysiologist for my entire professional career, I refer often to many of the features of the nervous system that Bell described or popularized. It has thus been a particular pleasure to write about him and to read or re-read many of his books and papers.

Michael J. Aminoff
San Francisco, California

ACKNOWLEDGMENTS

This work could not have been undertaken without the help of others. At the University of California, San Francisco, I am grateful especially to Mr. Aaron Daley for the patient help he gave me in compiling the bibliography of Bell and for tracking down often obscure reference material. Ms. Ariane Marcus and Ms. Ruth Gebrezghi were also helpful in obtaining reference material and preparing some of the illustrations. The staff of the library at the University of California, San Francisco, never failed to assist me, especially Ms. Azar Khatibi and Mr. Bazil Menezes. Dr. Imogen Hart of the Department of the History of Art at the University of California, Berkeley, kindly guided me to references concerning the influence on the art world of Charles Bell's *Essays on the Anatomy of Expression in Painting*, for which I am grateful.

Professor Emeritus Robert B. Layzer of the Department of Neurology at the University of California, San Francisco, read a penultimate copy of the entire manuscript, and I am particularly grateful to him for his helpful comments. He has been a colleague and friend for more than forty years. For some years, we had adjacent offices at the university, and I think of those years with great pleasure. I am also grateful to Professor Douglas S. Goodin, who read the penultimate version of two of the chapters and made useful suggestions. We have been friends, colleagues, and collaborators for more than thirty years, worked together in the laboratory, shared many happy times, and together watched our families grow.

Ms. Louise King, archivist at the Royal College of Surgeons of England, was kind enough to show me over the archives relating to Bell and to arrange for me to obtain illustrations for publication, and her colleague Mr. Bruce Simpson, curator of the Hunterian Museum, allowed me to reproduce a picture of the bust of Bell that is at the college. At the Wellcome Trust in London, Ms. Crestina Forcina, picture researcher, helped me to obtain pictures for publication from the vast collection of Wellcome Images, and Ms. Toni Hardy, archivist, helped to track down the paintings by Bell of the wounded from the battle of Waterloo, which are now in the possession of the Army Medical Services Museum at Keogh Barracks, Aldershot, Surrey (UK). Ms. Gail Anderson, curator of archives at that museum, kindly gave me permission to reproduce some of these paintings and also went to great trouble in attempting to locate Bell's original sketchbook and other notes, which unfortunately are missing. Ms. Marianne Smith, college librarian at the Royal College of Surgeons of Edinburgh, sent me information about the Corunna oil paintings at the college. Mr. George Richards, curatorial assistant at the University College London (UCL) Art Museum in London, provided information concerning the paintings held there. Ms. Kate Collins, librarian at the David M. Rubenstein Rare Book & Manuscript Library at Duke University, helped me obtain copies of Bell's illustrations in their collection. Mr. Jeremy Norman generously gave me access to his private collection of antiquarian books. Mr. Steven Wright, of UCL Library Services, was particularly helpful in giving me access to material held at UCL Special Collections at the college and in the National Archives at Kew, sending me digitized copies for my review. I am most grateful to them all.

I am also grateful for the help I received from my family. My father—Abraham S. Aminoff—died in 1994 but would have derived great pleasure from this book, given his love of history and his interest in the evolution of ideas. It saddens me that he will never see it. My wife, Jan, has been a wonderful companion to me over the forty years of our marriage, has taken great interest in my work, and has helped me in all my writing endeavors. She transcribed for me a number of the letters written by Charles Bell to Lord Brougham, and she read over the penultimate draft of much of the book. I dedicate this volume to her as a mark of my respect, love, and gratitude.

Our three children—Alexandra, a pediatric rheumatologist, Jonathan, a defense attorney working with the federal government, and Anthony, an attorney in the office of the Manhattan District Attorney—each read several chapters and made useful suggestions for improving the text. I appreciated their enthusiasm and support enormously.

Mr. Craig Panner, at Oxford University Press, has been a delight to work with, and I thank him for all his advice and help. His assistant Emily Samulski has also been most helpful, and I am grateful to her. Finally, I thank the production team who saw the book through to its publication, and especially the copy editor, Mr. Dan Hays, and production editor, Mr. A. Joseph Lurdu Antoine.

Michael J. Aminoff
San Francisco, California

1

AN INTRODUCTORY SNAPSHOT

The name of Charles Bell has been given to a nerve, a facial palsy, a clinical sign, an involuntary muscle spasm, a muscle, and a fundamental law of physiology, making it well known to physicians, surgeons, and medical students alike, even though they generally have little knowledge or appreciation of his actual, very solid accomplishments. Bell was revered by some contemporaries for his achievements: Philibert Joseph Roux, chief surgeon of a leading hospital in Paris, dismissed his class of medical students without a lecture immediately after having introduced Bell, with the words "C'est assez, Messieurs; vous avez vu Charles Bell." To others, however, Bell's name and brilliance were tarnished by charges of intellectual dishonesty and fraud, behavior that remains difficult to comprehend because Bell had so many real accomplishments to his credit. Thus, when historians focus on his scientific contributions, it is usually in relation to certain of his claims for scientific originality that were subsequently found to be of questionable validity (as discussed in Chapter 7), and his other contributions—important and insightful as they are (see Chapters 8 and 9)—are often quietly ignored. It is therefore important to set the record straight. The achievements and failings of Charles Bell deserve to be better known for—with time—they have faded into obscurity. In this chapter, therefore, Bell is introduced to the reader so that some idea can be gained about the breadth of his accomplishments before his life and work are considered in more detail.

THE SURGEON-SCIENTIST

Charles Bell, a Scottish surgeon–anatomist, lived in the last decades of the eighteenth century and first half of the nineteenth century, spending much of his professional life in London. A successful surgeon and clinician, he also had a real interest in the experimental underpinnings of medicine. Although his original publications in scientific journals are few in number, Bell added much to the sum of knowledge about the nervous system, and his contributions came to be widely recognized even as some of them became the subject of bitter controversy. Indeed, his discoveries concerning the nervous system were at one time likened to those of William Harvey on the circulation of blood, and his small pamphlet published privately in 1811, detailing his idea of the brain, has been described as the *Magna Carta* of neurology.

In that and subsequent publications, Bell suggested that the anterior and posterior nerve roots have different functions, showed that the nervous system has certain major divisions that are now taken for granted (the motor and sensory systems and the autonomic nervous system), and attempted to determine the central representation of motor and sensory nerves and thus to show that different parts of the brain have different functions. He also noted that individual peripheral nerves actually contain bundles of nerve fibers with different functions, that nerves conduct only in one direction, that sense organs are specialized to receive only one form of sensory stimulus, and that there is a sixth sense, namely a muscle sense. He even considered the basis for sensation, stressing that "perception is according to the part of the brain to which the nerve is attached." He suggested, in fact, new ways to look at—and to make sense of—the nervous system at a time when it was "puzzling in the last degree" and considered to be "inscrutable."

Bell lived during a time when an understanding of the nervous system—of its structure and the manner of its functioning—advanced remarkably to provide a solid foundation for all subsequent developments. He contributed significantly to this advance and received high honors in Britain for his scientific contributions, including the Royal Society Gold Medal (1829) and a knighthood (1831). However, a dispute that developed between Bell and François Magendie in France and Herbert Mayo in London regarding priority in discovering the separate functions of the nerve roots and certain cranial nerves led to charges of plagiarism by him and against him, making it uncertain for some years whether he was victim or villain.

THE CLINICIAN AND EDUCATOR

Bell's clinical acumen was notable. In addition to the facial palsy and its associated features named after him, he provided the first descriptions of several previously unrecognized neurological disorders that are now well known, although he did not always establish them as distinct entities (see Chapter 9). These included the first descriptions of muscular dystrophy, writer's cramp, numb chin syndrome, myotonia congenita, postinflammatory atlantoaxial dislocation, and several presentations of motor neuron disease. His reflections on various neurological phenomena, such as referred pain and reciprocal inhibition, are remarkable for their prescience.

During the Napoleonic Wars, Bell helped to treat the wounded and, based on his experiences, published *A Dissertation on Gun-Shot Wounds*, which served as a useful manual for military surgeons for years. He also used his artistic talent to create sketches and paintings of the wounded and the dead that are a dramatic contrast with the usual images and portraits of the glories of war encountered in museums and art galleries (see Chapter 5). They continue to be of interest as works of art and have provided much insight into the nature and treatment of war wounds during the early nineteenth century. For years before the advent of practical photography, they served as teaching aids in the training of doctors, both civilian and military.

Bell's *Essays on the Anatomy of Expression in Painting* (discussed in Chapter 4) helped to change the way art students are taught, described the anatomical basis of facial expressions, had a long-lasting effect in encouraging new approaches in the visual arts, initiated the scientific study of the physical expression of emotions, and led directly to the work of Charles Darwin—his former student—on facial expressions, published some sixty-six years later.

As an educator and educational reformer (see Chapter 11), Bell founded his own private medical school, subsequently took over the Great Windmill Street School of William Hunter, and eventually helped to found the University of London and the medical school of the Middlesex Hospital in London. Somewhat of an outsider in that great metropolis in the early nineteenth century, his views regarding the reform of education did not help him to make friends but were farsighted and important.

THE POLYMATH

Bell achieved much in the arts and sciences, in medicine and medical education. Some may wonder about the multiplicity of his interests and regard him as a dilettante rather than a person to be taken seriously. Others may question how he was able to succeed in so many different areas. His work must be viewed in the context of the man and the social, cultural, and even political times in which he lived. Bell was driven, ambitious, and hardworking, and he used his talents as an artist and illustrator to advance his career as a surgeon–anatomist and teacher. At the same time, he used his medical and scientific background to influence the visual arts by injecting realism into artistic representations of nature and human behavior.

His output as an author of clinical (as opposed to scientific) papers and textbooks was prodigious, with new works or new editions of his earlier works appearing in a steady stream, and this helped to extend his reputation from a local to a national level and then gave him an international standing. Indeed, many of his original scientific and clinical findings were published in this format rather than as papers in scientific journals. The times were also favorable—at least initially—to a man such as Bell. During the nineteenth century, learned men of different disciplines interacted with each other in a way that permitted the cross-fertilization of ideas, and there were fewer interruptions than there are today in the ordered life of the academic. Bell was clearly a learned surgeon–scientist, but he was also a talented painter and scholarly classicist. Times changed, however, and Bell was sidelined and eventually left behind, piously clinging to creationist beliefs that seemed increasingly dull and antiquated as new concepts emerged to counter them (see Chapter 10).

Bell lived a sober life, an academic bound up in his work and constantly leaning on his family for financial support. In his sixties, he returned to Scotland, an old man still holding to the creationist beliefs that framed his views as a surgeon–scientist. Although appointed to a prestigious chair at the University of Edinburgh, he felt redundant and unappreciated and became increasingly devoted to fishing and other leisurely pursuits. His creativity was diminished, he was worried about money, and he was sick with an ailing heart. He died in the arms of his wife while on his way to London in April 1842 and is buried in the old churchyard of a quiet English village.

2

IN THE BEGINNING

Early in the eighteenth century, Edinburgh, capital city of Scotland, consisted primarily of a main street running along an east–west axis between the Castle and Holyrood Palace, with narrow, dirty, overcrowded lanes branching off on either side. It was home to a university (founded in 1583 and the sixth oldest in Britain),[A] government offices, a cathedral, and various other ecclesiastical institutions. Its traditional industries were printing and distilling. The population numbered approximately thirty-five thousand, and there was close contact between the professional and working classes, who often shared the same alehouses and lived in the same tenement buildings, with the poor on the higher floors.

With the 1707 Act of Union between the kingdoms of England and Scotland, the parliament and many of the prosperous governing classes moved south to London. The economy initially declined, and the poor became increasingly unsettled by the high cost of living. Nevertheless, by mid-century, the city had expanded to the north and south of the castle, and the professional and business classes gradually moved from the medieval Old Town to the more spacious, elegant, single-family homes in the New Town. The professional class did particularly well: Lawyers prospered because Scottish law was distinct from English law, professors and medical men flourished because of the university and medical schools, and architects profited from the expansion of the city. Tradesmen also prospered because of the demands of an enlarging middle class, and the port of Leith, to the north, became important with the increasing commercial activity of the city and the arrival of trading vessels from all points of the globe. Prosperity also led to the establishment of various major banks and financial institutions. The last quarter of the eighteenth century and the first decade of the nineteenth century thus led to a remarkable expansion of the city, and the population reached over one-hundred thousand souls. Unlike Glasgow, the city did not become a major manufacturing or industrial center, but it gained an international reputation as an intellectual and cultural powerhouse, especially in philosophy, the arts and sciences, economics, and medicine.

[A] The University of Edinburgh was preceded in Britain by the universities of Oxford (1169 or earlier), Cambridge (1209), St. Andrews (1413), Glasgow (1451), and Aberdeen (1495). It is noteworthy that four of the six oldest British universities are thus in Scotland.

Medical education had a long history in Edinburgh, dating back to 1505 when the Incorporation of Barber Surgeons (which eventually became the Royal College of Surgeons of Edinburgh[B]) applied for its Charter as a Craft Guild and to the foundation of the Royal College of Physicians of Edinburgh in 1681. (The College received its Royal Charter from King Charles II in 1681 but had made three previous unsuccessful attempts in 1617, 1630, and 1656, which had met with resistance from the Edinburgh Incorporation of Surgeons, the Church,[C] the physicians and surgeons of Glasgow, and the universities, especially that of Aberdeen).[1,2] Botanical gardens, planted independently by surgeons and then by physicians in the latter half of the seventeenth century, were used primarily to teach medicinal botany, chemistry, and pharmacy (materia medica) to medical, surgical, and apothecary students and also to provide a stock of medicinal drugs.[3] The faculty of medicine at the University of Edinburgh was established in 1726, modeled on that at the Universities of Padua and Leiden (from which many of the influential local medical personalities of the period had graduated), and it is now one of the oldest in the English-speaking world. It soon attracted students from far and wide, boosting the economy of the city. In 1729, a small teaching hospital was opened in Edinburgh, receiving a Royal Charter from King George II in 1736. A new, purpose-built Royal Infirmary was opened in 1741 close to the university, providing access to material for clinical training, but larger facilities had to be built in the nineteenth century to accommodate the growing number of patients.

Although a complete medical education could be obtained at the University of Edinburgh in much of the eighteenth century, the extramural teaching of students—particularly in anatomy and surgery—grew ever more popular and led to a number of non-university medical schools each run by a group of teachers.[4] Many extramural teachers later became professors at the university, but the Incorporation continued teaching medicine independently and rented rooms to non-university lecturers for this purposes. In 1895, this practice became formalized into the School of Medicine of the Royal Colleges of Edinburgh, which closed in 1948.[4,5] Such, then, is the backdrop to the story of Charles Bell.

FAMILY BACKGROUND

Charles Bell was born in November 1774 in Fountainbridge,[D] a leafy suburb of Edinburgh. It was an uneasy year in Britain, a time of achievement but also

---

[B] It received its royal charter from George III in 1778.
[C] The Church believed that the establishment of such a college might restrict their privilege of awarding degrees. In Scotland at that time, the chancellors of St. Andrews, Glasgow, and Aberdeen Universities were either bishops or archbishops.
[D] As reported in the *Scottish Antiquary, or, Northern Notes and Queries*, Vol. 6, 1891, the original name for Fountainbridge was *Fauxbourgs* (suburbs). This was corrupted to *Foulbriggs*, and

of political turmoil. George III was king, and Lord North led a Tory government beset by troubles in the American colonies. The government adopted punitive policies after the citizens of Boston, disguised as Native Americans, took it on themselves in December 1773 to board the ships of the East India Company and dump their cargo of tea into the harbor. Harsh policies further stiffened American resistance, however, and—in September—led to the First Continental Congress in Philadelphia. Within a year, the war for independence would start, and a year later the Declaration of Independence would be adopted. The year 1774 was also the year that Clive of India died under rather mysterious circumstances, a year during which Captain James Cook continued his exploration of the South Pacific, and Joseph Priestley, the English chemist, discovered oxygen ("dephlogisticated air").

The Bell family were originally landowners who had reputedly been disinherited and deprived of their estate of Blacket House in Dumfriesshire by Oliver Cromwell because of their support for Charles I.[6] They moved to Glasgow and became successful merchants and then clergymen at a time when the Scottish clergy were influential members of the community rather than decorative relics of a past culture.

Charles was the youngest of eight children born to the Reverend William Bell (1704–1779) and his second wife, Margaret Morice. His paternal grandfather, the Reverend John Bell (1676–1708), was born in Glasgow but became Presbyterian minister at Gladsmuir in East Lothian and—with his natural eloquence—gave the sermon in Edinburgh cathedral on the death of King William III (William of Orange). Grandfather John married Janet, the daughter of a Major Learmont of Newholm, had five children, and died when only 32 years old. Charles' father, William, had little to do with his siblings, perhaps for religious reasons—he had left the Presbyterian Church (the Church of Scotland) for the Scottish Episcopal Church, which at the time was being repressed by the government because many of its members were supporters of the House of Stuart. For some years, William served as the Episcopalian minister at Doune, near Stirling, in Perthshire, initially earning only twenty-five pounds a year. In 1744, he resigned to become a colleague of Bishop Keith, in his Episcopal Meeting-house in Edinburgh, where he officiated as clergyman for thirty-five years and is said to have lived well.[6]

William's first wife, Lilias Grahame of Bowquaple, the daughter of one of the oldest families in Stirlingshire, had had two children, both of whom died in 1741, and she herself died in September 1750. William married again on 5 December

---

because *breig* or *brig* is the Scottish form of *bridge*, *foul* was converted to *fountain*, thus giving the more genteel Fountainbridge, despite there being neither water nor a bridge within any reasonable distance of it. Bell's date of birth is variously given as November 8 or 12.

1756, this time to Margaret Morice, the elder daughter of James Morice (sometimes spelled Morrice), Episcopalian clergyman of St. Andrews.[7] The father of his second wife had died soon after her birth, and Margaret and her sister Cecilia were raised by their maternal grandmother (also Margaret) and grandfather, Bishop Robert White, Episcopalian Bishop of Dunblane (1735–1743), Bishop of Fife (1743–1761), and Primus (presiding bishop) of the Scottish Episcopal Church (1757–1761).[8] Margaret was a good woman—pious and kind—and quite artistic, a talent that was apparently also possessed by her grandfather and that her sons Robert, John, and Charles also displayed.[9]

William saved money and bought several houses in Edinburgh; according to his nephew, the family moved to a small house and garden at Fountainbridge, near town, for the benefit of the children.[6] He is said to have noted in his diary that his purchase also "with the other house, rents in town, some money upon bond, I trust in God (whose good providence has enabled us to do so much), will be a moderate competency for my wife and children, should I be taken from them, and they become no burden to others."[6] This contrasts with statements made in the introductory chapter to Charles' published correspondence (presumably by his widow) that William left his family "very slenderly provided for," echoing comments made by both Charles and his brother George.[10] Perhaps, like so many others, William simply failed to realize the cost of raising a family; alternatively, Charles' widow may have emphasized the hardships faced by the family in order to give an added gloss to the subsequent career of William's sons.

Robert (1757–1816), the oldest son of William and Margaret, became a Writer (a solicitor or lawyer) to the Signet, an ancient branch of the legal profession in Scotland. The members of the society hold office under commission from the keeper of the Signet, an officer of the crown.[11] Writers to the Signet held special privileges regarding the drawing up of legal documents, being entitled originally to supervise use of the King's Signet, the private seal of the early kings of Scotland. Robert was admitted an advocate in 1812, was appointed lecturer on conveyancing by the Society of Writers to the Signet "to deliver annually a course of lectures on the theory and practice of conveyancing," and authored *A Dictionary of the Law of Scotland* and other legal works. He was a polished speaker, classics scholar, mathematician, accomplished musician, and skilled artist. As the eldest brother, he guided the education of the others after the death of their father in 1779.[9]

George, the next child, was born and died in 1759, as did William in 1763. John (1763–1820), the fourth child, was to be a major influence on Charles. He obtained the MD degree from Edinburgh University in 1779, became a fellow of the College of Surgeons seven years later, and soon after began teaching anatomy. He subsequently taught surgical anatomy in his own school in Surgeons'

Square,[E] where he performed dissections and established a museum. John was unusual for the time in suggesting that the practice of surgery must be based on a good understanding of anatomy,[12] and in this manner he was seen as competing with the professor of anatomy at the university, the illustrious Monro *secundus*.[F]

Among the pupils in John's large classes was his younger brother Charles, who became his apprentice and coauthor. Both brothers were articulate speakers and superb medical artists, a skill that aided their teachings and added luster to their books. John collected a number of anatomical specimens that Charles later took to William Hunter's museum at the Great Windmill Street School of Anatomy in London (see p. 25). In any event, the success of his school—and his habit of criticizing other anatomists (such as Monro)—led to increasing jealousy of John by his professional colleagues, and their hostility was not helped by his impulsiveness and volatility. Among his opponents was the eminent physician James Gregory, professor of the practice of medicine at the University of Edinburgh, who published and posted warnings to the students against John Bell and his lectures. An anonymous opponent, possibly Gregory but more likely John Barclay, a competitor with his own extramural school of anatomy in Edinburgh, even published in 1799 a satirical commentary under the pseudonym of Jonathan Dawplucker that seemed to praise a volume of John Bell's *Anatomy* (see p. 13) but in fact criticized both the work and its author as a plagiarist.[13,14] Gregory also urged that a small number of staff be appointed permanently to the infirmary rather than allowing members of the Royal College to act in rotation as surgeons there.[15,16] John was not among the six so appointed and thereupon gave up his teaching (in 1800) to concentrate on his private practice. He did, however, respond vigorously and in print to Gregory, complaining about the tone of his comments, defending the education and experience of the younger surgeons whose practice would be affected by a change in the appointment system, and supporting the system in which each surgeon had taken charge of patients at the infirmary for two months.[17,18] In any event, illness or the result

---

[E] Surgeons' Square was once the focal point of medical science and education in Edinburgh. It is now within the campus of the University of Edinburgh. Several private (extramural) schools were located there, as was the Surgeons' Hall with its theater for public dissections.

[F] Alexander Monro (1733–1817), Scottish anatomist and physician, was the second of three generations of physicians with the same name, hence *secundus*. He attended Edinburgh University when only twelve years old, subsequently became an assistant to his anatomist father, and eventually succeeded his father in the chair of anatomy. He described the lymphatic system, the foramen of Monro (an aperture between the lateral ventricles and the third ventricle of the brain, which had been described previously by others), and formulated the Monro–Kellie doctrine (that the volume inside the cranium is fixed such that any increase in volume of one constituent requires a decrease in volume of another). More brilliant than even his father, he was succeeded in turn by his son (Monro *tertius*), who failed to do justice to his illustrious pedigree and became a figure of fun.

of a riding accident led him eventually to retire, and he spent the last three years of his life as an invalid in Italy, studying art until he died of dropsy (now known as edema, usually from heart failure) in Rome in 1820.^G

The next two children of William and Margaret Bell, their daughter Margaret (1765–1832) and son James (1767–1830), never married and little is known of them. George Joseph, the seventh child, was born on 26 March 1770, and became professor of the law of Scotland in the University of Edinburgh, headed a commission in 1833 into Scottish bankruptcy law, and published several books, with his treatise *Principles of the Law of Scotland* becoming a standard textbook for law students. He died on 23 September 1843. He and Charles were remarkably close, and they kept in frequent contact with each other by letter when they were apart. These letters were published after the death of both brothers by Charles' widow, Marion (1788–1876), who outlived him by thirty-four years.[19]

CHARLES BELL AS A CHILD

Charles was born four and a half years after George Joseph (hereafter referred to simply as George) at suburban Fountainbridge, when his father was seventy years old. The family moved into Edinburgh some four years later because of his father's age and inability to travel to church in the city. A man of learning, his father's great delight was to sit by the window looking over the garden, surrounded by his books. He died just a few months after the move, and the family moved again, this time to the newly built George Street, which was still under development. There they lived in the upper part of a house, the rest of which belonged to his aunt, and managed to get by from the rent of their own houses, a widow's allowance from the Episcopalians, and "considerable aid" from their aunt Cecilia.

As Charles was only about five years old when his father died, and as little or no contact had occurred between his father and that side of the family, there was little paternal influence on the developing child. It was a matter of later regret to Charles that his only memory of his father was as a frail, ailing man:^H

---

^G John Bell spent his time in Italy studying art, making elegant notes on statues, paintings, and buildings as he proceeded. He admired the architecture and paintings that he saw, but he viewed the statuary with the critical, even mocking eyes of an expert anatomist. His wife published his notes five years after his death as *Observations on Italy*. The book had quite a success and was even published in Italian. Further details are provided by Avery H: John Bell's last tour. *Med Hist* 1964; 8: 69–77.

^H A number of the quotations are taken from Charles Bell's published correspondence with his brother George. These were published by his widow almost thirty years after his death, and the extent to which they have been edited is unknown.

His memory I always held in respect, but I knew him not as a man. I had no feature of his character to fix upon, and this I have often regretted. I have but one distinct recollection of him, in his dressing-gown, and that in his last illness my mother took me from betwixt his knees, being too lively, and insensible of his sufferings. . . .

When I look back to those days, my affection centres in my mother, and in my dear brother George. Yet Robert was most kind to me. I was his playfellow and pet.

My first recollections are as a little boy, by my mother's side. I recollect the dining-room, the view south, the mild affection with which she would stop her wheel, or point the letters (of my lesson) with her stocking-wire.[20]

Charles' mother was well educated, had a sweet disposition, and took great comfort from her two youngest sons. According to notes made by George, the education of both the brothers was stinted and they were—of necessity—self-taught. But Charles—he said—"had very solid judgment and a clear head . . . a great genius for drawing, and . . . was early destined for the profession of a surgeon under John."[21]

In his later years, Charles read in a short biographical note of himself that he was supposedly educated at the High School in Edinburgh,[22] a view that was later reiterated by others.[6] His response was emphatic:

I received no education but from my mother, neither reading, writing, cyphering, nor anything else.

My education was the example set me by my brothers. There has been, in my day, a good deal said about education, but they appear to me to put out of sight *example*, which is all in all. There was, in all the members of the family, a reliance on self, a true independence, and by imitation I obtained it.[23]

Despite these comments, Charles was in fact a pupil for four years (1784–1788) at Edinburgh High School, a famous school with many illustrious former students, some of whom were his contemporaries.[I] They included Walter Scott, Henry Brougham (afterwards Lord Brougham), and Francis Horner (see p. 66).[J] It is probable that his comments and recollections were colored by what and how

---

[I] The Royal High School is one of the oldest schools in the world, having been established nearly nine hundred years ago in 1128, and has enjoyed royal patronage since about 1590. In Bell's time, it was located on Infirmary Street, Edinburgh.

[J] Francis Horner was the younger brother of Leonard Horner, who became the first (and only) warden of the University of London when it was established. He is discussed in Chapter 11.

he was taught for he later confessed that his time there was one of "torture and humiliation" and failed to attract much attention:

> Anything demonstrative or mechanical, or tending to Natural Philosophy, I comprehended better than my companions; but the memory of verses or Latin rules, without intellectual comprehension of some principles, I was almost incapable of.
>
> This incapacity depressed me, and it was only when in professional education I found subjects more suited to my capacities that I began to respect myself, and favourably compare myself with my fellow students.[24]

One of the highlights of Charles' childhood was his visits from David Allen (1744–1796), the Scottish painter (sometimes known as the Scottish Hogarth), who habitually addressed him as "brother Brush."[K] Allen was short and rather ugly, with a coarse, pock-marked but animated face, and—when he wished—an engaging charm. He was patient, laudatory, and kind with Charles, encouraging him to copy drawings from the likes of Raphael and instilling in him a love of drawing for which he had an innate talent. In consequence, Charles came to work equally well when drawing in pencil, painting in watercolor or oil, making etchings and aquatints, or modeling in plaster or wax any cases of particular clinical or pathological interest.[25]

Charles was very close to his mother, and her death affected him deeply. Little is known about the cause, circumstances, or timing of her death except that he was about twenty years old:

> I hope I was a comfort to her. . . . She was my only teacher. . . .
>
> For twenty years of my life I had but one wish to gratify my mother, and to do something to alleviate what I saw her suffer.
>
> I suppose others feel as I did. It belongs to our nature to associate the being whom we love with our aspirations. When anything was proposed to be done, some fancy in my own mind, beyond my powers to attain, the question from childhood took always this fashion, "Could I not accomplish it were it to please her or save her?"
>
> No wonder that in losing her there was a long blank, an indifference to all accustomed objects. I hardly know how ambition was again produced in my mind.[26]

It is not clear to what Charles was alluding in speaking of her suffering. Perhaps it was the struggle of caring for a large family as a single parent or of having to manage with limited resources. The state of the family's finances remains

---

[K] Allen studied at the Foulis Academy in Glasgow and then spent several years in Rome, where he painted historical subjects and portraits as well as scenes of street life. On his return to Scotland, he enjoyed a successful career as both an artist (painter and etcher in aquatint) and a teacher.

unclear, however, for Charles went (despite his denials) to a highly regarded school,[L] had private art lessons from Allen, and was also tutored in French and Italian—implying that the budget was not as tight as he claimed.[6] It is also apparent from his comments that he became quite depressed after her death, withdrawing from his usual life and losing interest in events about him. Indeed, this was to become a pattern, with periods of brooding melancholy interrupting times of driven activity.

EARLY MEDICAL TRAINING AND WRITINGS

Charles attended the Edinburgh Medical School, training in anatomy and surgery under his brother John, and also teaching for him. Indeed, he became bound as an apprentice to John "in his calling of surgery and pharmacy" for five years starting on 2 April 1792, the indenture being witnessed by his brother George and also by Francis Bathie, John's apprentice at the time. He also attended the lectures of Monro *secundus*, professor of anatomy at the university (see footnote F, p. 9), as well as those of the philosopher Dugald Stewart (1753–1828) and of the mathematician and natural philosopher John Playfair (1748–1819). On 1 August 1792, he became a fellow of the Royal College of Surgeons of Edinburgh, prior to which he technically had been unqualified for the teaching role he had taken on at his brother's school. When John stopped teaching in 1799, Charles continued on his own until he left Scotland in 1804. He also assisted John in his operative work, and showed himself to be a speedy and innovative surgeon.[22] Together they wrote *The Anatomy of the Human Body*, published in four volumes between 1793 and 1804.[M] Charles did some of the illustrations for the first two volumes, which were written by John, and wrote and illustrated the last two volumes by himself. The third volume dealt with the anatomy of the nervous system and organs of special senses, whereas the final volume focused on the anatomy of the abdominal viscera, pelvic structures, and lymphatic system.

While a student, Charles also began to write *A System of Dissections, Explaining the Anatomy of the Human Body, the Manner of Displaying the Parts, and Their Varieties in Disease,* published in various parts between 1798 and 1803 and then combined into two volumes. This was his first independent work as an author, and he himself drew the illustrations, which were then printed from

---

[L] His school fees may have been paid for out of a trust fund, of which his brother Robert was trustee according to a letter by Robert's son, Charles, that was published in *Fraser's Magazine* in 1875.

[M] The first edition of Volume 1, by John Bell, was published in 1793 and was on the bones, muscles, and joints, with a second edition in 1797 when his Volume 2 (on the heart and arteries) was published. Volumes 3 and 4 by Charles Bell were published in London in 1802 and 1803 (Vol. 3, which was in two parts) and 1804 (Vol. 4).

plates made by various engravers. He emphasized the importance of dissection for learning anatomy for surgical purposes, stating in the preface,

> To study the details of anatomy, without having the parts before us, is pernicious: and a man, who has, by reading only, acquired a knowledge of names, and of the derivations of nerves and arteries, without at the same time being able to put his finger upon the body and tell what parts lie concealed, is more apt to be led astray, to hesitate and be timorous, than to be prompt and decisive in his conduct as a surgeon.[27]

Such a view would hardly be questioned today, although computerized techniques have become increasingly important in the study of human anatomy. Charles went on to state explicitly the purpose of his own book:

> The object of this work is to serve as an assistant to the student in acquiring a knowledge of Practical Anatomy; in gaining a local memory of the parts; in learning to trace them upon the dead subject, and to be able to represent them to his own mind upon the living body. This being my object, the method to be pursued is obvious: to give a short detail of the anatomy; to show how the parts are to be laid open, and how they are to be distinguished in dissection, or avoided in an operation; to explain the consequence of each part to the great functions of the body, and to mark the diseases to which it is liable.[27]

He provided detailed instructions on such topics as how to dissect bodily parts (for example, by dissecting muscle along the direction of its fibers), the injection of arteries or veins, the preparation and staining of bone and marrow specimens and of the solid viscera (such as the heart, lungs, kidneys, and spleen), and the methods for preserving tissues. During dissection, he emphasized the utility of practical but often neglected procedures such as the introduction of probes into ducts (such as the salivary ducts) and of catheters (fine tubes) into a body organ.[28] In addition to describing regional anatomy, he also summarized the anatomy and function of individual muscles and of the major blood vessels and their branches. These volumes thus provided a standard practical guide for students of surgical anatomy and met with wide acceptance.

Two other important works were published by Charles during this period and attest to his artistic talent. His *The Anatomy of the Brain, Explained in a Series of Engravings* (1802) contained twelve beautiful stipple-engraved anatomical plates (a number of which were hand-colored) accompanied by detailed explanatory notes.[29] This was followed in 1803 by *A Series of Engravings, Explaining the Course of the Nerves*, which contained nine copper-engraved plates,[30] all made from his original drawings.

Charles owed much to his family. Although fatherless since the age of five years, his mother was devoted to him, and his elder brother, Robert, became of necessity a father figure, guiding his education and upbringing. His brother John taught Charles anatomy and surgery, and he served as a professional role model not only in these areas but also as an illustrator, writer, educator, and—perhaps unfortunately—as an outspoken critic of those with whom he disagreed. And George, just four years his senior, became his close confidant and friend, a relationship that was maintained for years by letter when they were apart. The young Charles was talented but had a modest and simple manner and an engaging personality that made him many friends. Nevertheless, life was becoming more difficult in Edinburgh, where neither John nor Charles had gained positions at the infirmary when the appointment system there was changed in 1800. Academically, too, opportunities were lacking.[9] The chair of anatomy had been filled for another lifetime (by Monro *tertius*), and two new chairs of surgery had recently been created (in part because of the inadequacies of Monro) and filled by others.[N] It seemed also that the ill-feeling toward John would be a continuing hindrance to him, a bar to his future advancement. Even his offer to pay the infirmary and donate his anatomical collection had failed to gain him permission to make notes and drawings as bodies were dissected there.[25] Ambitious and confident of his ability but unwilling to push himself forward (a characteristic that seemingly changed as he became older and better known), Charles began to look elsewhere for opportunities to advance.

In 1804, George advised him to go to London to gain further experience during the winter and to remain there if he found a suitable position. It was a desperate measure, for he had no friends in London and—indeed—did not know a soul. Nevertheless, on November 23, Charles left the Scottish for the British capital, moving from the current to the future center of British medical education.

REFERENCES

1. Kaufman MH: Early history of the Royal College of Physicians of Edinburgh. *Res Medica* 2005; 268: 49–53.
2. Gairdner J: Sketch of the early history of the medical profession in Edinburgh. *Edinb Med J* 1864; 9: 681–701.
3. Doyle D: Edinburgh doctors and their physic gardens. *J R Coll Physicians Edinb* 2008; 38: 361–367.
4. Craig WS: *History of the Royal College of Physicians of Edinburgh.* pp. 337–352. Blackwell: Oxford, 1976.

[N] A chair of clinical surgery (1803) founded at the university entirely through his effort went to the surgeon James Russell, a former president of the College of Surgeons, and a chair of surgery (1804) at the College of Surgeons was filled by John Thomson, an eminent pathologist.

5. Guthrie D: *Extramural Medical Education in Edinburgh and the School of Medicine of the Royal Colleges.* Livingstone: Edinburgh, 1965.
6. Bell C (nephew of Sir Charles Bell): Letter to the editor. *Fraser's Magazine (n.s.)* 1875; 12: 129–130.
7. Scotland, Marriages, 1561–1910, index, *FamilySearch* (https://familysearch.org/ark:/61903/1:1:XYMV-KHW; accessed 17 May 2015), Morice in entry for William Bell and Margaret Morrice, 05 Dec 1756; citing Edinburgh Parish, Edinburgh, Midlothian, Scotland, reference; FHL microfilm 1,066,689, 103,046.
8. Kaufman MH: Genealogy of John and Charles Bell: Their relationship with the children of Charles Shaw of Ayr. *J Med Biogr* 2005; 13: 218–224.
9. Struthers J: Historical sketch of the Edinburgh Anatomical School. *Edinb Med J* 1866; 12: 431–446.
10. Bell M (?): Note concerning William Bell's wife and family. p. 8. In *Letters of Sir Charles Bell, K.H., F.R.S. L. & E: Selected from His Correspondence with His Brother George Joseph Bell.* Murray: London, 1870.
11. Society of Writers to Her Majesty's Signet: *A History of the Society of Writers to Her Majesty's Signet with a List of the Members of the Society from 1594 to 1890 and an Abstract of the Minutes.* pp. CVII– CXXVI. Constable: Edinburgh, 1890.
12. Kaufman MH: John Bell (1763–1820), the "father" of surgical anatomy. *J Med Biogr* 2005; 13: 73–81.
13. Walls EW: John Bell, 1763–1820. *Med Hist* 1964; 8: 63–69.
14. Dawplucker J: *Remarks on Mr. John Bell's Anatomy of the Heart and Arteries.* Robinson: London, 1799.
15. Gregory J: *Memorial to the Managers of the Royal Infirmary.* Murray & Cochrane: Edinburgh, 1800.
16. Gregory J: *Additional Memorial to the Managers of the Royal Infirmary.* Murray & Cochrane: Edinburgh, 1803.
17. Bell J: *Answer for the Junior Members of the Royal College of Surgeons, of Edinburgh, to the Memorial of Dr. James Gregory.* P. Hill: Edinburgh, 1800.
18. Bell J: *Letters on Professional Conduct and Manners; on the Education of a Surgeon, and the Duties and Qualifications of a Physician. Addressed to James Gregory, M.D. Professor of the Practice of Medicine in the University of Edinburgh.* John Moir: Edinburgh, 1810.
19. Bell C: *Letters of Sir Charles Bell, K.H., F.R.S. L. & E: Selected from his correspondence with his brother George Joseph Bell.* Murray: London, 1870.
20. Bell C: Recollections of his mother. pp. 10–11. In *Letters of Sir Charles Bell, K.H., F.R.S. L. & E: Selected from His Correspondence with His Brother George Joseph Bell.* Murray: London, 1870.
21. Bell G: Memoranda of my life by George Bell. pp. 9–10. In *Letters of Sir Charles Bell, K.H., F.R.S. L. & E: Selected from His Correspondence with His Brother George Joseph Bell.* Murray: London, 1870.
22. Pettigrew TJ: Sir Charles Bell, K.H., F.R.S. L. & E. Professor of surgery in the University of Edinburgh, &c. &c. &c. pp. 1–22. In *Biographical Memoirs of the Most Celebrated Physicians, Surgeons, etc. etc. Who Have Contributed to the Advancement of Medical Science.* Whittaker: London, 1839.

23. Bell C: Recollections of his mother. p. 10. In *Letters of Sir Charles Bell, K.H., F.R.S. L. & E: Selected from His Correspondence with His Brother George Joseph Bell*. Murray: London, 1870.
24. Bell C: Education at the high school. p. 13. In *Letters of Sir Charles Bell, K.H., F.R.S. L. & E: Selected from His Correspondence with His Brother George Joseph Bell*. Murray: London, 1870.
25. Corson ER: Sir Charles Bell: The man and his work. *Bull Johns Hopkins Hosp* 1910; 21: 171–182.
26. Bell C: Recollections of his mother. pp. 11–12. In *Letters of Sir Charles Bell, K.H., F.R.S. L. & E: Selected from His Correspondence with His Brother George Joseph Bell*. Murray: London, 1870.
27. Bell C: Preface. pp. v–vi. In *A System of Dissections, Explaining the Anatomy of the Human Body, the Manner of Displaying the Parts, and Their Varieties in Disease*. Mundell & Son: Edinburgh, 1798.
28. Bell C: Introduction, pp. vii–xiv. In *A System of Dissections, Explaining the Anatomy of the Human Body, the Manner of Displaying the Parts, and Their Varieties in Disease*. Mundell & Son: Edinburgh, 1798.
29. Bell C: *The Anatomy of the Brain, Explained in a Series of Engravings*. Longman and Rees: London, 1802.
30. Bell C: *A Series of Engravings, Explaining the Course of the Nerves*. Longman and Rees: London, 1803.

3

LONDON AND THE GREAT WINDMILL STREET SCHOOL

Now thirty years old, Charles Bell arrived in London five days after leaving Edinburgh on the horse-drawn mail coach, traveling along what is now the Great North Road and resting at York to view the Minster. After a stop at The Bull and Mouth Inn, the great coaching-inn on Aldersgate Street, he stayed at The London Coffee House on Ludgate Hill—a place much favored by visiting businessmen and Americans, and which is said to have been frequented by Benjamin Franklin—before moving into lodgings at 10 Fludyer Street, Westminster.[A] As Bell described it, the only other person at the lodging-house was an old lady, once of some distinction but now in reduced circumstances, who never left her room. There was a maid—a staid, decent person—who provided for him: "steaks, veal, pigeon-pie, rabbits, fowls, regularly a new-laid egg to supper, ham, cheese, &c."[1]

Bell had no personal friends in London, but his name was already well known because of his published works. He wasted little time in contacting a number of prominent surgeons and anatomists, people at the top of their profession, hoping to find work, perhaps as an assistant. Among those with whom he met were Astley Cooper,[B] Matthew Baillie,[C] William Blizard,[D] William Lynn,[E] and

---

[A] Fludyer Street, Westminster, exists no longer—having been replaced by government buildings—but apparently ran parallel to Downing Street and opened on to St. James's Park. It was named after Sir Samuel Fludyer who, as Lord Mayor of London in 1761, entertained George III and Queen Charlotte at the Guildhall.

[B] Sir Astley Cooper (1768–1841) was surgeon to Guy's Hospital, London, and an eminent anatomist who made major contributions to otology, the treatment of abdominal aneurysms (by tying the aorta), diseases of the breasts and testicles, and the surgery of hernia. He was serjeant-surgeon (i.e., part of the medical household of the sovereign) to King George IV, King William IV, and Queen Victoria. His statue stands in St. Paul's Cathedral, London.

[C] Matthew Baillie (1761–1823), anatomist and physician extraordinary to King George III, was the nephew of the Hunter brothers (surgeons and anatomists).

[D] Sir William Blizard (1743–1835), English surgeon, was founder of the medical school at the London Hospital (now the Royal London Hospital), which was the first medical school attached to a hospital in England. It is said that he introduced the practice of "walking the wards," now a standard feature of medical education.

[E] William Lynn (1753–1837), surgeon at the Westminster Hospital, London, was said to have had the largest operating practice in the metropolis.

John Abernethy,[F] two of whom (Blizard and Abernethy) went on to found medical schools at the London and St. Bartholomew's Hospitals in the capital, and another (Baillie) who took over the private medical school founded by William Hunter at Great Windmill Street. (Bell himself would go on later to establish a medical school at the Middlesex Hospital, another major London hospital, as discussed in Chapter 11.) He also met with James Wilson, the anatomist, whom he eventually joined at the Great Windmill Street School in 1812 (Fig. 3.1).[G] He breakfasted with Sir Joseph Banks ("a kingly figure of an old man with a blazing star upon his breast")[H] and went on to meet with Anthony Carlisle,[I] surgeon at the Westminster Hospital, whom he described as "a man of some abilities, but having the greatest conceit of himself I ever knew a man to possess."[2] Bell had hoped that Carlisle—himself an artist with an interest in the relationship of the arts to anatomy—might be of some help to him but instead Carlisle spoke cuttingly that they preferred to manufacture "their own raw material" and indicated that if there had been difficulties for Bell in Edinburgh, there would be more in London.[2]

Bell remained in lodgings for the next year, lecturing and assisting at surgical operations (which he had occasional opportunities to sketch), performing anatomical studies at the Westminster Hospital, teaching anatomy to art students, drawing from life at the Royal Academy, and completing his book on the *Anatomy of Expression*, an unfinished draft of which he had brought with him from Edinburgh.[3] He gave drafts to several colleagues, including Banks, who was impressed. But he felt that many opportunities simply were not open to him, and at times would despair of the future. As he put it, "These days of unhappiness and suffering tended greatly to fortify me, so that nothing afterwards could come amiss, nothing but death could bring me to a condition of suffering such as I then endured."[4]

He was, of necessity, careful with money and—after buying new clothes—once declared that he looked "primitive and parson-like."[5] When depressed, he would impulsively resolve to return to Edinburgh and his family there, only

---

[F] John Abernethy (1764–1831), English surgeon, was a popular lecturer, a quaint eccentric, and founder of the medical school of St. Bartholomew's Hospital, London.

[G] James Wilson (1765–1821), Scottish anatomist, studied under John Hunter and eventually took over his position at the Great Windmill Street medical school. Among his pupils was James McGrigor (see Chapter 5).

[H] Sir Joseph Banks (1743–1820), English botanist, was president of the Royal Society for more than forty years. He sailed with Captain Cook on his first great voyage to the South Pacific, Australia, and New Zealand, and he was largely responsible for establishing Kew Gardens, near London, as one of the most important botanical gardens in the world.

[I] Sir Anthony Carlisle (1768–1840), English surgeon, is said to have discovered electrolysis by passing a voltaic current through water, decomposing it to hydrogen and oxygen.

Figure 3.1 (Top) A lecture at the Hunterian Anatomy School, Great Windmill Street, London. Watercolor, 1830, by Robert Blemmel Schnebbelie. (Bottom left) The home and anatomy school of William Hunter in Great Windmill Street. (Bottom right) James Wilson, surgeon and teacher of anatomy at the school, by Thomas Joseph Pettigrew. Wilson bought the school, and Bell—in turn—subsequently paid him two thousand pounds for it, with the proviso that Wilson could continue to live and teach there. (Courtesy of the Wellcome Library, London.)

to change his mind the next day as things looked up. The social distractions of London, with its theater and opera, and increasing recognition and friendship with a few professional colleagues cheered him and helped put off the decision to leave.

Social gaffes sometimes helped. On one occasion, he arrived for dinner at the house of Sir William Blizard to find he had come on the wrong week—he was nevertheless invited to join the family meal and was swiftly put at ease. "[T]his visit, of course, made us ten times more intimate than the formal dinner could have done with eighteen or twenty black-coated fellows of doctors."[6] He also became close to William Lynn, who had a large surgical practice at the Westminster Hospital, assisting him with occasional surgical cases, going out with him to the theater, or joining him at his country cottage, where he would read or write while Lynn gardened, or they would walk together in the paddocks.[J] He dined with Astley Cooper several times and agreed to make some drawings for him, although declining an invitation to stay with him. He became close, also, to John Abernethy, and agreed to make drawings for one of his assistants, a comparative anatomist at St. Bartholomew's Hospital named James Macartney, who was writing "some little thing" on the arteries (that might have competed with Bell's own book).

He also had his detractors, who viewed him as an ambitious and scheming opportunist. One of them told Lynn that Bell "was a sharp, insinuating young man" and that "before he was aware, I [Bell] would have him out of his hospital!!"[7]

BELL'S OWN SCHOOL OF ANATOMY

As no appointments at established institutions were forthcoming, Bell came to realize that he needed to set up his own school rather than continue to hope for a position elsewhere. In September 1805, he rented a large but dilapidated house at 10-11 Leicester Street (Leicester Square) where he remained until 1811, when—shortly before his marriage—he moved to Soho Square, a more fashionable area. The house (now demolished) had been built in the late seventeenth century and was occupied successively for brief periods by the Earls of Manchester and Stamford, then by Lord Newborough, the Duchess of Leeds, and Arthur Onslow, Speaker of the House of Commons, who remained there from 1727 to 1753.[8] The neighborhood was one that was favored at one time or another by a number of distinguished men such as Hogarth, Joshua Reynolds, and John Hunter, the anatomist.

[J] Bell dedicated the second edition of his *A System of Operative Surgery Founded on the Basis of Anatomy* to Lynn as a public expression of the warmth of his friendship and respect for his professional talents.

The house—with its three storeys, eight garrets in the attic, and extensive grounds at the back—was large enough that he could lecture there and take in pupils.[K] The main floor consisted of, at the front, "... 'the Great room Forward', for it had 'five sash Windows glaz'd with the best Crown Glass.' Behind it lay a bedchamber with two windows, and a dressing-room with one. Next to these were the great staircase and the back stairs..."[9] On the ground floor, the front was divided between a large room functioning as an entrance hall and opening onto the street and the front parlor. At the back was the withdrawing room, with an adjacent dressing room, and an unheated back closet.[9] Among the second-floor rooms was a front bedroom and dressing room, a second bedroom, spare room, housekeeper's bedroom, and another dressing room. The fireplaces in the ground and first-floor rooms were of marble. The basement probably included a powder room, pantry, laundry, steward's hall, larders, and kitchen.

George had come down from Scotland to visit Charles in July 1805, and both brothers liked the house and thought that together they could manage the rent. Bell's surveyor viewed it more gloomily, ruminating especially about the state of the roof. As Bell remembered it,

> When I went with my surveyor to examine it, I was somewhat appalled by his account; he was a great John Bull rough fellow. Leaning out of the window, and observing the walls out of their perpendicular, he said in a coarse, familiar manner, "Sir, you had better have nine bastard children than this house over your head."[10]

Nevertheless, Bell—after he had found a roofer to maintain it at a relatively low cost—went ahead with a repairing lease before any misgivings could hold him back. With suitable premises and with a widening professional circle and growing reputation, Bell began to create his own school. Classes began at the end of September. In January 1806, he started to give public lectures in a little theater he had set up, taught surgical anatomy in dissecting rooms that were open from eight in the morning to ten o'clock at night, and admitted painters into his house, where they drew in his great room the skulls and skeletons from his collection. The first few days were slow as students trickled in, but at least they came. And his school proved such a success that by the end of the year he was able to write to George,

> You are anxious about my money matters. That is nonsense. I have got a great deal of money from my pupils, from my class, from Longman's [his publisher], from my patients. ... I have got essentials, but not comforts.[11]

[K] A few pupils (house pupils) would live in the house and work closely with him, helping to prepare and maintain his museum, helping with dissections, and learning by apprenticeship.

His *System of Operative Surgery*,[L] a work on general surgery from an anatomical perspective and based on his personal experience, was soon to be published by Longman's, with twenty-one full-page engraved plates after his own drawings. It included sections on orthopedics, trauma, neurosurgery, ophthalmology, otorhinolaryngology, and abdominal, head and neck, genitourinary, and vascular surgery. He advocated careful dissection based on an understanding of anatomical principles rather than the less deliberate surgery of the period. With the money he expected from this and his other books, as well as the hoped-for income from his school, he was becoming more optimistic about his finances. Troubled by difficulty in retaining staff, however, he subsequently learned that they were frightened off because his house was thought to be haunted by the ghost of a beautiful young lady.

Bell used the introductory chapter from his as-yet unpublished book, *The Anatomy of Expression*, as a successful introduction to a series of lectures to painters—six originally signed up for the course but the number soon doubled, at two guineas each.[M] He also took on pupils whom he tutored, and within a few weeks his surgical pupils had brought in eighty-two pounds and his painters twenty-five pounds. In February, he earned his first fee for a clinical consultation. And so he became busy and soon—in his own words—was in "a perpetual hurry from one thing and another, and *I like it!*" [emphasis in original].[12]

PRIVATE SCHOOLS OF ANATOMY

Charles Bell was by no means unusual in establishing a private school of anatomy and surgery in London. Several schools already existed and played an important part in medical education in the early nineteenth century, although in many instances cadavers for dissection were obtained illegally. Entry into medical practice then typically involved completion of an apprenticeship, but medical qualifications could—if desired—be obtained by examination from certain British universities (Oxford and Cambridge in England and five Scottish universities) or Colleges of Physicians or Surgeons (two in England and three in Scotland).[N] Membership of the Society of Apothecaries was by examination for those wishing to practice in London.

[L] The work subsequently went through two editions in Britain and into American, Italian, and German editions. His dissertation on gunshot wounds (see Chapter 5) was published as an appendix to the second (1814) edition as well as in a free-standing form.

[M] A guinea was one pound and one shilling.

[N] A medical degree from Oxford or Cambridge primarily involved studying the writings of physicians such as Hippocrates from ancient times and then defending in Latin a thesis before the professor of medicine.

Medical practitioners were physicians or, far more commonly, surgeons, apothecaries, or surgeon–apothecaries (typically a surgeon who then also practiced as an apothecary without additional qualification). In London, the surgeons had separated from the Company of Barber-Surgeons in 1745 to form the Company of Surgeons (which in 1800 became the Royal College of Surgeons),[13] and private courses in anatomy and dissection—previously a monopoly of the combined company—became permitted and led to the establishment of many private schools. Surgical skills and expertise were passed on primarily by example to the pupils or assistants of an established surgeon.

THE GREAT WINDMILL STREET (HUNTERIAN) SCHOOL

Among the private schools, the most famous was that of William Hunter (1718–1783), who trained in Glasgow and then in London and was an illustrious obstetrician, physician to Queen Charlotte, and professor of anatomy at the Royal Academy of Arts from 1769 until 1772, thereafter retaining strong connections to the artistic world. He was also an accomplished and enthusiastic teacher, initially assisted by his brother John (1728–1793),[o] although the brothers became estranged over which of them had discovered the circulation of the placenta. William set up his own school in London and later planned to establish in London a national school of anatomy with a museum to house his large and valuable collection of normal and pathological specimens, as well as more diverse biological and anthropological material.[14] He applied to the government for a grant of land on which to build, but nothing materialized.[p] He therefore proceeded at his own cost but somewhat less grandly, purchasing and demolishing a dwelling at 16 Great Windmill Street, near Piccadilly, to replace it with a large house of his own—with lecture theater, dissecting room, and museum—where he established his new school in 1769.[15,16] He organized systematic courses and laced his lectures with anecdotes to make them interesting. Beginners were discouraged from taking notes but urged instead simply to grasp the principles under discussion. Strangers were forbidden entry, and visitors were not allowed to attend lectures on the organs of generation and the gravid uterus. Demonstrations preceded dissections.[16]

---

[o] John Hunter subsequently set up his own anatomy school in London and accumulated nearly fourteen thousand preparations of more than five hundred different species of plants and animals, which formed the basis of the Hunterian Museum at the Royal College of Surgeons of England. He was appointed surgeon to St. George's Hospital and, in 1776, surgeon-extraordinary to King George III. Many of the specimens in the Hunterian Museum were lost or damaged by bombing in World War II, but the collection remains a very impressive one.

[p] The site that Hunter was seeking is part of that on which the National Gallery in London now stands.

A succession of fine anatomists joined the school, some becoming partners of Hunter or breaking away to found their own school.[17] On Hunter's death, William Cruikshank (1745–1800)[Q] and Matthew Baillie (see footnote C, p. 19) inherited the freehold of the premises, the medical school, and the use of the museum for thirty years. The museum and some eight thousand pounds were eventually to go to the University of Glasgow, and the museum was sent to Scotland by sea in 1807. James Wilson (1765–1821) began at the school as Baillie's demonstrator and then lectured there, purchasing the freehold of the premises and living in the house until his death.

The reputation of the school began to decline after Hunter's death. Bell may have begun special evening lectures there (on surgery)[18] in 1811 and, early in 1812, Wilson offered to sell him the school for ten thousand pounds, a sum that Bell could not possibly afford. A deal was concluded, however, allowing Bell to pay two thousand pounds with the proviso that Wilson could continue to live and teach at the school. The money may have come, at least in part, from a small dowry that he received when he married (see p. 79),[3] but it was "all my money to the last penny" and "eighteen pence more."[19] Thus, Bell took over Hunter's School and with it the collection of specimens that Wilson had by then accumulated—combined with his own, he now had a magnificent anatomical museum. He gave his first lecture at his new school on 7 October 1812, to between eighty and one hundred pupils. As he wrote to George, he had now fulfilled his greatest ambition with regard to teaching.[20] But there were heavy demands on his time, and he lectured only after much preparation, for the reputation of the school depended on the talents and enthusiasm of its teachers.

Bell lectured for two hours daily on anatomy, physiology, pathology, and surgery, all of which he regarded as aspects of a more general course on anatomy. In addition, he lectured on surgery for three evenings each week.[18] Unlike many of the private schools, the Great Windmill Street School also provided teaching in physiology and pathology (as part of its anatomy course), midwifery, and diseases of women and children.[18] Bell was assisted in his lectures, demonstrations, and preparation of specimens at the school by several other anatomist-surgeons, including the Shaw brothers, who were to play a large part in his life. One of their sisters, Marion, became Bell's wife and another married his brother George, as discussed in Chapter 6.

John Shaw (1792–1827), the son of Charles Shaw, clerk of the county of Ayr (Scotland), had been sent to London in 1807, when aged fifteen, to become a pupil of Bell. At the Great Windmill Street School he came to act as superintendent of

---

[Q] Cruikshank's own collection of anatomical specimens was bought by the Russian government and taken to St. Petersburg.

the dissecting room and, after Wilson's death, became a lecturer as well. He performed much of the experimental work that led to Bell's published discoveries on the nervous system, and he also had a major role in forming Bell's anatomical museum. Shaw accompanied Bell to Brussels immediately after the battle of Waterloo to care for the wounded and study the effect of gunshot wounds, as discussed in Chapter 5. Two years before his death, he was elected surgeon to the Middlesex Hospital, where he proved to be an outstanding orthopedic and spinal specialist. When John died, his younger brother, Alexander (1804–1890), took over his role at Bell's school, and he subsequently had a long and illustrious career at the Middlesex Hospital. Among Alexander's published works are various accounts of Bell's researches on the nervous system.

When John Shaw indicated in early 1824 his wish to withdraw from lecturing at the school in order to spend more time practicing surgery, this almost certainly helped Bell in deciding to give up the Hunterian school. In November, he decided to get rid of his large museum, which he sold the following year to the Edinburgh College of Surgeons for three thousand pounds, a purchase that the College authorized by a vote of thirty-five to fourteen in March 1825 and which was completed in July.

Bell and John Shaw conducted the last session of the school in 1825–1826. The timetable and cost of studying there are recorded:[21] Lectures by Bell and Shaw on anatomy, physiology, and pathology were given daily at 2 P.M. with a charge of five guineas for the first, four guineas for the second, and three guineas for the third course, there being no charge for attending lectures thereafter. Courses were given in the winter and spring, with the spring course concluding with lectures on the operations of surgery. The dissection rooms were open from 9 A.M. until 2 P.M. from October to mid-April, where Shaw gave regular and full demonstrations of the dissected parts, explained the application of anatomy to surgery, demonstrated operative methods on dead bodies, and taught the art of making anatomical preparations. The cost of the demonstrations was three guineas for each of four courses. Finally, Bell gave evening courses on surgery for three guineas.[21]

That summer, Bell disposed of the school to a former student, Herbert Mayo (1796–1852), and to anatomist Caesar Hawkins (1798–1884) and in 1826—when he parted from the school—he agreed to pay back four hundred pounds of the purchase price if he taught at any competing school of anatomy.[R] Bell went on to teach at the newly formed University of London, which became University College when King's College was founded in 1830. Hawkins moved as surgeon

---

[R] The foundation stone of University College (the original University of London) was laid in 1827, and Bell's acceptance of its chair of physiology (which he regarded as a branch of anatomy) therefore cost him the four hundred pounds.

Figure 3.2 (Top) Charles Bell, aged thirty years, was very anxious to be appointed professor of anatomy at the Royal Academy. (Courtesy of the US National Library of Medicine.) (Bottom) Print showing Sir Joshua Reynolds (holding his ear trumpet) at the Royal Academy. Standing to his left, William Hunter, the first professor of anatomy at the academy, is directing the arrangement of a male model. Johann Zoffany, 1783. (Courtesy of the Wellcome Library, London.)

to St. George's Hospital in 1830, and Mayo eventually transferred his lectures to King's, where he became the first professor of anatomy. So the Hunterian school at Great Windmill Street came to an end.[19]

ANATOMY AND ART

Bell completed *The Anatomy of Expression* at the house on Leicester Street, publishing it in 1806. The book emphasized to painters the value of studying anatomy and of illustrating emotional expression in their work, but it was also an important contribution to the medical literature. This work—to be discussed in detail in Chapter 4—and his close contact with artists whom he taught at the Leicester Street school led him perhaps to consider a future as an artist and illustrator and so to aspire to the position of professor of anatomy at the Royal Academy of Art (Fig. 3.2).[22] This institution had been incorporated in 1768 by King George III with Sir Joshua Reynolds as its first president and William Hunter as the first professor of anatomy. The professor was elected by the forty members of the academy. John Sheldon succeeded Hunter but had been in poor health for several years and unable to deliver all his planned lectures. When he died in 1808, letters were received from Joshua Brookes, Anthony Carlisle, and Charles Bell, offering themselves as candidates for the vacant position. Many favored Charles Bell, who was well-liked by artists and anatomists alike.[23] Although Bell actively sought the nomination and asked for the help of his friends, perhaps surprisingly it was Anthony Carlisle who was appointed, as discussed further in Chapter 4, in which Bell's interests in the arts are also examined.

REFERENCES

1. Bell C: Letter to George dated 19 April 1805. pp. 42–44. In *Letters of Sir Charles Bell, K.H., F.R.S. L. & E: Selected from His Correspondence with His Brother George Joseph Bell*. Murray: London, 1870.
2. Bell C: Letter to his brother dated 30 November 1804. pp. 21–22. In *Letters of Sir Charles Bell, K.H., F.R.S. L. & E: Selected from His Correspondence with His Brother George Joseph Bell*. Murray: London, 1870.
3. Anon: Sir Charles Bell, K.G.H., 1774–1842. Br J Surg 1920; 8: 389–391.
4. Bell C: Note on his early struggles. pp. 68–69. In *Letters of Sir Charles Bell, K.H., F.R.S. L. & E: Selected from His Correspondence with His Brother George Joseph Bell*. Murray: London, 1870.
5. Bell C: Letter to George dated 4 December 1804. p. 23. In *Letters of Sir Charles Bell, K.H., F.R.S. L. & E: Selected from His Correspondence with His Brother George Joseph Bell*. Murray: London, 1870.
6. Bell C: Letter to his brother dated simply as "Saturday, after supper." pp. 28–29. In *Letters of Sir Charles Bell, K.H., F.R.S. L. & E: Selected from His Correspondence with His Brother George Joseph Bell*. Murray: London, 1870.

7. Bell C: Letter to George dated 1 January 1805. pp. 30–32. In *Letters of Sir Charles Bell, K.H., F.R.S. L. & E: Selected from His Correspondence with His Brother George Joseph Bell*. Murray: London, 1870.
8. Hibbert C, Weinreb B, Keay J, Keay J (eds.): Leicester Street. p. 481. In *The London Encyclopaedia*, 3rd edition. Macmillan: London, 2008.
9. Anon: Leicester Square, North Side, and Lisle Street Area: Leicester Estate, Leicester Street. pp. 476–480. In Sheppard FHW (ed.): *Survey of London: Volumes 33 and 34. St. Anne Soho*. London County Council: London, 1966.
10. Bell C: Note on early struggles. pp. 69–70. In *Letters of Sir Charles Bell, K.H., F.R.S. L. & E: Selected from His Correspondence with His Brother George Joseph Bell*. Murray: London, 1870.
11. Bell C: Letter to George dated 20 December 1806. pp. 85–86. In *Letters of Sir Charles Bell, K.H., F.R.S. L. & E: Selected from His Correspondence with His Brother George Joseph Bell*. Murray: London, 1870.
12. Bell C: Letter to George dated 11 February 1806. pp. 66–67. In *Letters of Sir Charles Bell, K.H., F.R.S. L. & E: Selected from His Correspondence with His Brother George Joseph Bell*. Murray: London, 1870.
13. Jackson B: Barber–Surgeons. *J Med Biogr* 2008; 16: 65.
14. Thomson SC: The Great Windmill Street School. *Bull Hist Med* 1942; 12: 377–391.
15. Hawkins C: London teachers of anatomy: A retrospect. *Lancet* 1884; 124: 535–536.
16. Power D'A: The medical institutions of London. *Br Med J* 1895; 1: 1388–1391.
17. Thomson SC: The surgeon–anatomists of Great Windmill Street School. *Bull Soc Med Hist Chicago* 1942; 5: 55–75.
18. Crosse MV: *A Surgeon in the Early Nineteenth Century: The Life and Times of John Green Crosse M.D., F.R.C.S., F.R.S., 1790-1850*. pp. 33–34. Livingstone: Edinburgh, 1968.
19. Struthers J: Historical sketches of the Edinburgh Anatomical School. Sir Charles Bell. *Edinb Med J* 1866; 12: 436–446.
20. Bell C: Letter to George dated 7 October 1812. p. 203. In *Letters of Sir Charles Bell, K.H., F.R.S. L. & E: Selected from His Correspondence with His Brother George Joseph Bell*. Murray: London, 1870.
21. Anon: Medical, surgical, and anatomical schools. *Lancet* 1825; 5: 87–88.
22. Loudon ISL: Sir Charles Bell and the anatomy of expression. *Br Med J* 1982; 285: 1794-1796.
23. Darlington AC: *The Royal Academy of Arts and Its Anatomical Teachings; With an Examination of the Art Anatomy Practices During the Eighteenth and Early Nineteenth Centuries in Britain*. PhD thesis, University of London, 1990.

4

ANATOMY OF THE EXPRESSION OF EMOTIONS

Renaissance artists, particularly Leonardo da Vinci, Michelangelo,[A] and Raphael, performed dissections to advance their anatomical knowledge and improve their work.[1] Indeed, Leonardo probably intended to publish a treatise on anatomy, recognizing the mechanics involved in the human form, but after his death in 1519 the significance of his drawings remained unrecognized for centuries. It was Vesalius, therefore, who came to be regarded as the father of modern human anatomy based on his 1543 publication of *De Humani Corporis Fabrica*, which was probably illustrated by various artists. Later masters produced *écorchés*, skinless studies of the muscles—sometimes peeled away—to study the nude figure in a different way. Despite the symbiosis of art and anatomy, it was only in the latter part of the eighteenth century that anatomy was formally introduced into a British educational institution for the arts, when William Hunter was appointed the first professor of anatomy at the newly created Royal Academy of Arts in 1768.[1]

Hunter was an entertaining teacher, using anecdotes to add interest to even the most banal of points, and engaging his students in anatomical studies. He probably obtained his cadavers illegally as it was not until the Anatomy Act was passed in 1832 that the dissection of donated cadavers and those of unclaimed paupers by medical practitioners, anatomy teachers, and certain students was permitted.[B] Previously, only the corpses of executed murderers could be used for this purpose, and an illegal trade in corpses had developed to make up the shortfall. This may have been the reason that the rules of the Academy stipulated that all visiting lecturers had to submit their

---

[A] Michelangelo (1475–1564) dissected cadavers as a teenager and later hoped to publish an anatomy book for artists. However, he destroyed or burned most of his anatomical drawings before his death, and only a few survive. In his painting of the ceiling of the Sistine Chapel are various anatomical representations. The mantle of God in his *Separation of Land and Water* has been likened to a bisected right kidney (Eknoyan G: Michelangelo: Art, anatomy, and the kidney. *Kidney Int* 2000; 57: 1190–1201). Similarly, his *Creation of Adam* contains within it an image of the brain, and his *Separation of Light from Darkness* contains a view of the brainstem (Suk I, Tamargo RJ: Concealed neuroanatomy in Michelangelo's *Separation of Light from Darkness* in the Sistine Chapel. *Neurosurgery* 2010; 66: 851–861).

[B] Bell was one of a number of prominent surgeon–anatomists who pressed for changes to the law, leading ultimately to the Anatomy Act.

lectures for approval, except for the professor of anatomy.[1] It would seem that a blind eye was turned to the practice by the establishment.

Bell was well known for having produced many important self-illustrated works on anatomy and surgery, and he developed his own medical school with a museum (now held at the Royal College of Surgeons in Edinburgh) that contained an important collection of dissections preserved or copied in wax. He strongly believed that modeling, sculpting, and painting helped surgeons to both understand anatomical relations and improve their hand–eye coordination. He himself clearly had great manual skills given his work as surgeon and anatomist. His anatomical drawings were visually pleasing while also demonstrating the anatomical relationships between different parts of the body. He was also well known as a teacher of anatomy to artists in his private school, having instructed many artists, including Edwin Landseer and his brothers,[c] Benjamin Haydon,[D] and David Wilkie.[E]

WAX ANATOMICAL MODELS

The study of anatomy in the sixteenth and seventeenth centuries was limited by difficulty in obtaining cadavers and the unpleasantness associated with dissecting cadavers that had not been preserved.[2,3] The introduction of wax anatomical models obviated these problems, and their use permitted a three-dimensional appreciation of anatomical relationships and served as an alternative to dissected human specimens. A useful account of their development has been provided elsewhere.[2] Some models were created by casting, but others were sculpted, and the use of colored wax added realism.[3] These models were accurate anatomically, and some even displayed early signs of decomposition.

The credit for leading the way in creating such realistic models belongs to Gaetano Giulio Zumbo (or Zummo) (1656–1701), a Sicilian abbot who spent the

---

[c] Sir Edwin Henry Landseer (1802–1873), artist, renowned for his paintings of animals, was Queen Victoria's favorite painter. He sculpted the four bronze lions at the base of Nelson's Column in Trafalgar Square, London; the bronze was from the cannons of French and Spanish ships captured at Trafalgar. As a young man, he dissected animals in order to appreciate their muscular and skeletal structure. He declined election to the presidency of the Royal Academy.

[D] Benjamin Robert Haydon (1786–1846), English painter and writer, was interested in anatomy and taught a number of prominent painters. He did not himself achieve real fame as a painter, but his autobiographical writings have fascinated many by their insight into and revelations of his disturbed mind and the information provided about many of his contemporaries. He was imprisoned for bankruptcy and eventually committed suicide. William Wordsworth, John Keats, and Elizabeth Barrett Browning all dedicated sonnets to him or wrote about him.

[E] Sir David Wilkie (1785–1841) was a Scottish painter. Returning from a trip to the Holy Land in 1841, he died at sea, an event commemorated in a famous painting by Turner.

last year of his life in Paris. He worked with surgeons in Genova and Paris to create quite remarkable models, and the technique spread to other cities and countries. The wax models created in Bologna and then in Florence were especially notable, with collections of normal or pathological anatomical parts, obstetrical conditions, and botanical specimens being made for Italian or foreign universities, especially as aids to the teaching of medical students. In England, cadavers were available more readily than in many other parts of Europe, and it was only after their use became regulated that the art of creating anatomical models really developed in Britain.[2] The models created were especially realistic but were also perhaps less artistic and refined, more cadaveric, and more likely to shock than those from Italy, which were often made to look alive, even sensuous, and comfortable.

Bell spent a lot of time on his wax models, and he discovered a technique that allowed them to retain their color in its original freshness, which added to their utility.[4] As Sydney Smith—the writer, cleric, and fashionable lecturer—wrote to Francis Jeffrey,[F] a friend since childhood of Bell and his brother George,

> I hope to see more of your friend Bell. He is modest, amiable, and full of zeal and enterprise in his profession. I could not have conceived that anything could be so perfect and beautiful as his wax models. I saw one to-day which was quite the Apollo Belvedere of morbid anatomy.[4,G]

THE ANATOMY OF EXPRESSION

Bell had arrived in London with various essays that he put together and published in 1806 as *Essays on the Anatomy of Expression in Painting*, with various plates. His philosophy was clearly stated in the preamble to the book

> The painter must not be satisfied merely to copy and represent what he sees; he must cultivate this talent of imitation, merely as bestowing those facilities which are to give scope to the exertions of his genius, as the instruments and means only which he is to employ

---

[F] Francis Jeffrey (1773–1850) was a Scottish lawyer, judge, and literary critic. He was one of the founders of the *Edinburgh Review* and its editor for twenty-six years.

[G] Given Bell's skill as an anatomist and modeler, it is not surprising that when the French anatomist Louis Auzoux (1797–1880), who had devised a new modeling technique, planned to visit Britain in 1832, the naturalist Baron Georges Cuvier (1769–1832), wrote to Bell, introducing Auzoux and commending his technique. The new method utilized papier-mâché, which was easier to work with than wax; it produced models that were less fragile and could be dismantled to reveal how the body was put together. The visit was well timed, as it occurred while the Anatomy Act was under discussion in parliament. Cuvier's original, signed letter to Bell, dated 24 January 1832, was made available to me at the Alain Brieux antiquarian bookshop, 48 rue Jacob, Paris, France.

for communicating his thoughts, and presenting to others the creations of his fancy. It is by his creative powers alone that he can become truly a painter; and for these he is to trust to original genius, cultivated and enriched by a scrutinizing observation of nature. Till he has acquired a poet's eye for nature, and can seize with intuitive quickness the appearances of passion, and all the effects produced upon the body by the operations of the mind, he has not raised himself above the mechanism of his art, nor does he rank with the poet or the historian.[5]

The book, written in terms that nonmedical readers could understand, gained an immediate popularity, went through several reprintings and seven editions during which it was revised and expanded, and became a favorite of Queen Charlotte and, later, Queen Victoria. The second edition[H] was published in 1824, and a third and later editions were published posthumously, with the third edition (1844) being perhaps the most influential. The dedication of the second edition to his brother George Joseph Bell is especially touching:

> The reprinting of this volume recalls the time when it was written, when we studied together, before the serious pursuits of life were begun. I inscribe it to you, as the just object of my pride as well as of my affection; . . . so that . . . [others] may learn how much brotherly attachment adds to the value of life.[6]

Unlike many books of the period, it was well illustrated with etchings and plates to emphasize the points being made, although in later editions the text expanded at the expense of the illustrations. Many of the illustrations were by Bell himself, but some (unattributed) may have been contributed by others, including David Wilkie. In his own drawings and paintings, Bell often depicted sections and fragments of the body to focus the viewer's attention on the structures being discussed.

Bell wrote to his brother, with tongue in cheek,

> The Queen spoke to Maton of it . . . with great admiration, and begged him to convey her thanks to me. One of the ladies of the bed chamber told Maton[I] the Queen was reading it

---

[H] Renamed for the second edition as *Essays on the Anatomy and Philosophy of Expression*, it was published by Murray. Bell had written to him in August 1821 explaining that Longmans—who had published the first edition on the basis of divided profits—wished to cut it down, whereas he wished to improve on it and have a volume of taste and judgment. Further details are provided by Smiles S: *A Publisher and His Friends: Memoir and Correspondence of the Late Murray, with an Account of the Origin and Progress of the House, 1768–1843*. Volume II, p. 119. Murray: London, 1891. Subsequent editions were titled *Essays on the Anatomy and Philosophy of Expression as Connected with the Fine Arts*.
[I] William George Maton (1774–1835), English physician and a man of great charm, was appointed physician extraordinary to Queen Charlotte, wife of King George III, in 1816, and he was subsequently closely associated with the royal family.

for two hours last night. Oh, happiness in the extreme, that I should ever write anything fit to be dirtied by her snuffy fingers![7]

Its success, coming as it did from Bell's artistic ability, literary skill, and profound knowledge of anatomy and physiology, was not surprising.

Bell believed that God's design was visible in nature, and a theme of the book is natural theology (see Chapter 10).[8] He also believed that the human figure (rather than landscape painting) should be the prime concern of the artist, and the human body has, in fact, had a major place in Western art for many centuries. He considered that knowledge of anatomy would help artists to understand the muscle actions involved in various expressions and other activities and further their interpretive skills.[8] Anatomy taught artists how to "see" nature and thus how to portray action. Without an anatomical framework, artists would not be able to portray the human body appropriately and thereby do justice to God's creation. He critically appraised contemporary education in the arts, which required that students draw from other drawings and models rather than cadavers,[9] and indicated the role that anatomy should fulfill:

> The display of muscular action in the human figure is but momentary, and cannot be retained and fixed for the imitation of the artist. The effect produced upon the surface of the body and limbs by the action of the muscles, the swelling and receding of the fleshy parts, and that drawing of the sinews or tendons, which accompanies exertion, or change of posture, cannot be observed with sufficient accuracy, unless the artist is able to class the muscles engaged in the operation; and unless he have some other guide than the mere surface presents, which may enable him to recollect the varying form. . . .
>
> In natural action there always is a consent and symmetry in every part. When a man clenches his fist in passion, the other arm does not lie in elegant relaxation.[10]

In his *Essays*, he discussed the changes in the skull and face that occur from infancy to old age (Fig. 4.1). As the infant and child grows to adulthood, the frontal and maxillary sinuses enlarge, the jaws increase in size and length (with the chin becoming more prominent), and the cheekbone and nose come to project more so that "a new character is given to the whole countenance." The oval, elongated shape of the infant skull also changes but differs in different races. For painters, the changes in form of the frontal bone are especially important: the character of the whole head depends to a greater or lesser extent "on the contour of the forehead, the ridges of the temples, the prominences formed by the cavities in this bone, and, lastly, the arch of the orbit." In the elderly, the face again changes: For example, the teeth fall out, the gum tissues waste away, the front of the upper and lower jaws move closer to each other, "the chin and nose approach," and the lips fall in (Fig. 4.2). These and other age-related changes that

Figure 4.1 Changes in the skull and face with age. Compared to the adult, the infant has an elongated head, flat forehead, small nose, small and short jaw, and small neck because of the prominent occiput. (From Bell C: *Essays on the Anatomy and Philosophy of Expression*, 2nd edition. p. 181. Murray: London, 1824.)

he characterized need to be taken into account—Bell urged—in representing the human form rather than following "mechanical rules" for drawing the face.

Bell argued that humans had certain facial muscles not present in other animals, that these had been designed by God to expand the range of human expression, and that these were responsible for the great versatility and variety of human expressions and for the activity required to modify the lips and cheek and thereby modulate the voice.[11] In another essay, he compared the facial muscles of humans—and the large repertoire of human facial expressions—with the limited facial expressions of other animals. He discussed specific expressions (which he related to specific emotions) and their anatomical basis. The emotions were expressed not by changes in a single feature but by changes involving a number of muscles and thus much of the face. He pointed out that with the exception of the few muscles that move the lower jaw, the facial muscles are attached at one end to bone and at the other to skin or elastic tissues (the moveable tissues) of the face.

There are, for example, different types of smile, and these convey different meanings depending on the context and particularly on the activity of associated muscles. Such smiles include the "placid smile of benignity; the

Figure 4.2 Old age. With loss of the teeth in the aged, the alveoli (supporting tissues) waste away and nothing remains but the narrow base of the jaw. The jaws thus approach closer to each other at the front of the mouth; the chin and nose approach, and the mouth is too small for the tongue, the lips fall in, and the speech becomes inarticulate. (From Bell C: *Essays on the Anatomy of Expression in Painting.* p. 29. Longman, Hurst, Rees, and Orme: London, 1806.)

contemptuous arching of the lower lip; the smile of sorrow; the simper of conceit; the distorted smile of the drunken man, when the eyes with difficulty perform their office; the leer, &c. ...." A smile is produced by activation of the same muscles as in laughter, but to a lesser degree, and Bell analyzed the involved muscles in these activities. Other emotions were analyzed similarly, in anatomical detail:

> In joy the eyebrow is raised moderately, but without any angularity; the forehead is smooth; the eye full, lively, and sparkling; the nostril is moderately inflated, and a smile is on the lips. In all the exhilarating emotions, the eyebrow, the eyelids, the nostril, and the angle of the mouth are raised. ...
>
> In rage the features are unsteady, the eyeballs are seen largely; they roll and are inflated. The front is alternately knit and raised in furrows by the motion of the eyebrows; the nostrils are inflated to the utmost; the lips are swelled, and being drawn, open the corners of the mouth. The action of the muscles is strongly marked. The whole visage is sometimes

pale, sometimes inflated, dark, and almost livid; the words are delivered strongly through the fixed teeth;[12]

The muscles around the eyes, he felt, were especially important in generating a lively appearance, and large eyes were essential to beauty. In the second edition of his book, he indicated his particular interest in the expression of infants as—he believed—they showed especial purity and exemplified the manner in which facial expressions can generate as well as express the emotions they represent.[13]

> The expression of pain in an infant is extraordinary in force and caricature; the expression of laughter is pure in the highest possible degree, as indicating unalloyed pleasure, and will relax by sympathy even the iron features of a stranger. Here the rudiments of expression ought to be studied, for in after life they cease to have the pure and simple source which they have in infancy.[14]

He studied the facial expressions of the insane, as well as those of politicians in parliament and actresses such as Sarah Siddons on the stage.[15] He depicted extreme emotions such as rage as indicated by facial expression and bodily posture, or the blank expressions associated with the stillness, apathy, and inactivity of extreme melancholia. Bell held that understanding the individual limb or facial muscles in isolation was not adequate, and that artists should study the parts in relation to the whole to give full expression of the predominant emotions of individual subjects.[16]

In his book, Bell emphasized that the object of painting was to capture not just beauty of form but also the power of action and of human expression. To him, both expression and bodily actions and postures were an important aspect of beauty. Although a portrait should resemble the sitter, that in itself was insufficient and the artist must also endeavor to capture the person behind the mask. Others, artists or academicians, held a quite different view, namely that expression detracted from beauty.

Bell remarked that the varying expressions of the face indicate the condition of the mind.[17] He also stressed that "the mind is dependent on the frame of the body"—for example, mimicking the looks and gestures of someone who is angry or afraid can lead to those very emotions—and that philosophers had tended to overlook the relation between the mind and the condition of the body.[18] Certain states of mind, he pointed out, have a very wide influence on the body. As an example, he considered the expression of terror (Fig. 4.3):

> We can readily conceive why a man stands with eyes intently fixed on the object of his fears, the eyebrows elevated to the utmost, and the eye largely uncovered; or why, with

A

B

Figure 4.3 (A) Fear. The eyeball is largely uncovered; the eyes staring; the eyebrows raised. Breathing is abnormal, with a gasping in the throat, distension of the nostrils, convulsive opening of the mouth, and dropping of the jaw; the lips nearly conceal the teeth but allow the tongue to be seen. (Courtesy of the Wellcome Library, London.) (B) Terror. The eye is bewildered, the eyebrows are knit together and their inner part is turned up, and distracted thought, anxiety, and alarm are indicated by the expression. The cheek is a little elevated and the muscles about the mouth are all activated. It is sometimes suggested that this drawing was by David Wilkie, but the evidence is unclear. (From Bell C: *Essays on the Anatomy of Expression in Painting*. pp. 142 and 146. Longman, Hurst, Rees, and Orme: London, 1806.)

hesitating and bewildered steps, his eyes are rapidly and wildly in search of something. In this we only perceive the intent application of his mind to the object of his apprehensions—its direct influence on the outward organ. But observe him further: There is a spasm on his breast, he cannot breathe freely, the chest is elevated, the muscles of his neck and shoulders are in action, his breathing is short and rapid, there is a gasping and a convulsive motion of his lips, a tremor on his hollow cheek, a gulping and catching of his throat; and why does his heart knock at his ribs, while yet there is no force of circulation?—for his lips and cheeks are ashy pale.[19]

Although Bell's point of view concerning the beauty of expression followed that of various German scholars, philosophers, and artists,[20] he was among the first to advance this approach in Britain and to encourage its wide acceptance through his school and writings. His acceptance of the widely held view that the ancients assimilated the finest parts from a number of individuals of the same age and gender to produce figures representing perfection was derived from a study of ancient sculptures.[20] To Bell, however, ideal figures were produced by the artist's grasp of nature, preceded by anatomical study to maximize knowledge of natural structure. Bell believed in the subjectivity of the aesthetic response, in contrast to the many others who sought some quality or form that resided in an object and that in itself generated esthetic responses.[20]

He criticized painters who gave to animals the facial expressions of humans (unless the animals were simply emblematic), because they miscast the functional beauty of animal forms.[20] In like manner, he criticized painters who could not distinguish between posture and action and thus whose work showed "tameness" because they had neglected the play of the muscles. He viewed the Parthenon sculptures[J] privately late in 1807 when few had yet seen them and found them pleasing because they recorded so faithfully the natural appearance of animals and humans.[20] The collection was not yet complete and they were placed in "something of an accidental confusion" as he wrote to his brother, but they were striking:

> There is a piece which, if it had been entire, would have surpassed all other remains of antiquity. It may be Neptune rising from the sea, from an irregular surface as of water; the arms are just emerging and half-floating, the neck and shoulders are seen, but the head and face is broken off; there is strength and beauty united in the neck, shoulders, and arms, and there is something in the inclination of the figure implying buoyancy, and as if the broad chest were rising on the surface of the water.[21]

The sense of motion, of activity, and of purpose that emerges gives this piece to Bell an especial attraction of its own.

Bell believed in the necessity of a complete understanding of the anatomy and physiology of emotions so that the artist need not simply copy nature but

---

[J] The Parthenon Marbles, a collection of classical Greek marble sculptures and architectural pieces from the Parthenon and adjacent buildings on the Acropolis of Athens, were brought to London by Lord Elgin in the early nineteenth century, subsequently purchased by the British government, and are on display in the British Museum. Their acquisition—of uncertain legality—continues to lead to protests and demands that they be returned to Greece.

can be more creative without confounding the barriers and restrictions that exist in nature.[20] And he considered a child to be the most pleasing—and thus the most beautiful—object in the world because the "natural form of a child is the only species of beauty so perfect in character and expression, that it cannot be excelled by art, nor receive addition by the adoption of an ideal form."[22]

The book was a great success. John Flaxman (1755–1826), professor of sculpture at the Royal Academy, believed it had done more for the arts than any other contemporary thing.[13] Henry Fuseli (1741–1825), professor of painting and keeper at the academy, also praised it and sent Bell three beautiful engravings from his pictures. Charles Darwin was stimulated by it to pursue the study of expressions, which he detailed in his own book (p. 44).

With the increase in student numbers in both art and medical education during the eighteenth and early nineteenth centuries, a new market force had been created:

> The influences that Sir Joshua Reynolds, William Hunter, Charles Bell and John Marshall had on both the artistic and medical communities cannot be underestimated for not only did they produce cultural icons of the body but they were instrumental in bringing anatomical education to artists who would have otherwise been denied it.[16]

John Sheldon (1752–1808), the professor of anatomy at the Royal Academy of Art, had been in failing health with mental illness (possibly bipolar disease) for several years.[K] When he died in 1808, the academy set about appointing a replacement. As Bell wrote to George, while starting—even before Sheldon had died—to seek influence for gaining the position, "The voters are the Academy, forty in number. Those who have influence are his Majesty, the Royal Family, nobility, and gentry: all who ever had their faces drawn. "[23] The candidates were Joshua Brookes, Anthony Carlisle, and Charles Bell.[16] Given his background, experience, and teachings, his interest in the visual arts and talent as a painter, and the enthusiasm and eagerness with which he encouraged his students,[24] it is not surprising that Bell wished to be appointed. Such an appointment would have brought him prestige, influence, power, and financial reward. He seemed the most obvious choice, was a favorite with artists and anatomists alike, and

---

[K] Sheldon had lectured on anatomy at the Great Windmill Street School under William Hunter, whom he subsequently succeeded as professor of anatomy at the Royal Academy. His anatomical work was focused on the lymphatic system. He is reputed incorrectly to have been the first Englishman to ascend in a balloon—his first attempt ended when the tethered balloon caught fire, and his second when the balloon failed to ascend.

did some quiet canvassing. Astley Cooper wrote to Sir William Beechey[L] about the forthcoming election:

> Bell, of all the men I know, beyond all comparison merits the situation of Professor to the Royal Academy. He adds to a very extended knowledge of anatomy a perfect acquaintance with the principles of painting, and I feel the strongest conviction that if he is elected, he will do infinite credit to himself, and be an invaluable acquisition to the Royal Academicians. If I had the ear of the king, I should tell him he ought not to vote for any other person. Yours, &c, A. C.[25]

Wilkie and Haydon also put in a word to Beechey and others on behalf of Bell.[26] Bell wrote to George, in an undated letter,

> I have been spending the evening with Mr. and Mrs. Winn. Lord Headley has been with Sir Jos. Banks [president of the Royal Society], and Sir J. says I am the person best entitled to the situation, and if his name can be of any service he shall be happy, or call on those he can influence.[27]

And again, "You cannot think, my dear George, how much it goes against my feelings to canvass the business of this lectureship among those men."[28]

Nevertheless, Anthony Carlisle, although not the overwhelming first choice with either the artists at the academy or their medical colleagues, was appointed. Thomas Lawrence—a portrait painter who, in 1820, became president of the academy—was one who had opposed Bell and who wrote after the election,

> I am quite convinced that the academy has escaped a mischief in not electing Mr. Bell, whose whole conduct has betrayed such total want of temper, modesty, and judgment as to show him to be a very unfit man for our society.[29]

To what Lawrence was referring is unclear—perhaps it was simply a personal antipathy, perhaps Bell's canvassing, perhaps his less refined manner; but certainly, Bell's candidacy was not helped by his critique of the "tameness" of certain portrait painters. In any event, the choice of Carlisle was curious because, in an essay in *The Artist*, Carlisle had earlier expressed the opinion that a minute knowledge of anatomy was not necessary to the painter and sculptor of history.[30] "Anatomy is capable of affording instruction to the mind," he wrote, "but, except in works of natural history, it should not be seen in the penciling of an artist." Not surprisingly, then, the *Annals of the Fine Arts*, in an attack on the academy,

---

[L] Sir William Beechey (1753–1839) was an English portrait-painter and member of the Royal Academy.

commented, "Mr. Charles Bell, was more adequate for the professorship of anatomy, because he had written on the benefits of its use in painting, than Mr. Carlisle, who had written that it was no use at all."[31]

It seems that it was the portrait painters "who elected Carlisle."[16] The greater social acceptability of Carlisle may have been a deciding factor for many. Both men were talented and skilled anatomist–surgeons and highly regarded in the medical profession, but Carlisle was also well known in London's literary circles and as a patron of the arts. The conservatism of the academicians probably led them to elect professors of similar sociopolitical standing, and this social advantage would have helped Carlisle to win the appointment. Bell was more extreme than Carlisle even in medical politics, favoring a new approach to medical education rather than adapting the existing establishment, and therefore—it may have been thought—would not have fitted in so well. He was critical of the way anatomy was being taught and also of painters who did not know anatomy. Carlisle had good, carefully cultivated, connections that could only have been an asset to him. On 28 December 1808, Carlisle was officially elected the professor of anatomy.[16] He was swift to make his mark as professor, giving sensational lectures that were extravagantly theatrical and delivered "in full court dress with bagwig and cocked hat."[32] He is said on one occasion to have passed around a human brain and heart on dinner plates while speaking of the emotions,[32] perhaps in a dig at Bell.[M] When he resigned in 1824 after holding the appointment for sixteen years, the council of the academy voted to pay fifty pounds for a commemorative salver to be presented to him.

Bell applied again for the professorship when Carlisle resigned. The president on this occasion was Thomas Lawrence—the self-same Lawrence who had been opposed to Bell sixteen years earlier, considering him unacceptable for social and political reasons.[8,28] The chair again went to another candidate, this time to Joseph Henry Green (1791–1863), who is also known for his appointment as the literary executor of Samuel Taylor Coleridge. Like Carlisle his predecessor, Green was an establishment man who opposed the "lecture bazaars" of nonconformist anatomists and did not feel it necessary to reject the existing system in order to reform it.

[M] Exhibitions of groups of nude figures placed in motion for the display of particular muscles were popular attractions at Carlisle's lectures. In 1821, Carlisle engaged a squad of the Life Guards (the senior regiment in the British Army) to show how the muscles were exercised when using the broadsword. A crowd gathered and—despite the presence of officers from Bow Street—stormed the Royal Academy buildings, with some of the more daring spirits climbing on the roof of the Lecture Room and pushing through the windows in the hope of seeing the show, which Carlisle was forced to cancel. More details are provided by Whitley WT: Turner as a lecturer. *Burlington Mag* 1912–1913; 22: 257–258.

But what of Bell's famous book? It was to have a significant impact in both the scientific world and among artists.

BELL'S *ESSAYS* AND THE SCIENTIFIC STUDY OF EXPRESSION

Bell had focused on the face and its expressions, and he analyzed the expression of the emotions in anatomical terms, distinguishing between different emotions. In his second and later editions, he had also emphasized the role of the nervous system in generating the expressions of emotions. These were remarkable achievements. As a young medical student in Edinburgh, Charles Darwin (1809–1882) had been a student of Bell, had studied his book, was much impressed by his analyses of facial expressions, and later utilized Bell's anatomical descriptions in his own writings. Over the years, however, he found it more difficult to accept the creationist views of Bell and others or the idea that species were immutable, and he became increasingly opposed to the view that humans had certain facial muscles not present in other animals to expand the range of human expression.[N] As discussed in Chapter 10, the intellectual climate had changed, and theories of evolution had begun to displace the creationist views that had once prevailed.

As did Bell, Darwin examined in detail the actions of the facial muscles in the expression of emotion, but—in contrast to Bell—he also attempted to understand why particular expressions were associated with specific emotions.[33] He concluded that these actions were the same in humans all over the world (and thus supported the belief for "the several races being descended from a single parent-stock") and were not restricted to humans, and that precursors of the muscles were present, for example, in nonhuman primates. In other words, Darwin believed that such facial expressions must have had a common evolutionary origin, and that they are innate and not learned (unlike gestures).

Darwin's book, *The Expression of the Emotions in Man and Animals*,[34] published in 1872, detailed his findings and conclusions. It was intended originally to be a single chapter but, as the evidence accumulated, Darwin decided that the work merited a volume of its own; it was completed in four months, as a sequel to his *The Descent of Man* (1871). The book, one of the first with photographs, was a best-seller. Darwin expressed his special indebtedness to Bell

---

[N] In 1826, Darwin had joined the Plinian Club (founded in 1823, closed in 1841), where students could discuss papers on natural history. A topic of much discussion was Bell's book with its creationist views. Speakers railed against it, claiming that the lower animals possessed all the faculties of the human mind and that humans did not differ from other animals in their muscles of facial expression. Such discussions must surely have influenced the young Darwin.

(to whose third—1844—edition he referred constantly), as well as to Guillaume Duchenne (de Boulogne) for his work on the electrical stimulation of the facial muscles, and to a variety of other colleagues, including those who had helped him gather information from distant lands. Duchenne (1806–1875), a French neurologist and electrophysiologist, had applied his electrical stimulation techniques to studying the physiology of what he regarded as a divinely inspired universal language of facial expression, drawing heavily on Bell's work to locate and identify the facial muscles.º

When considering the different emotions in humans, Darwin studied not only different ethnic groups but also—as did Bell (Figs. 4.4 and 4.5)—the expressions of the mentally ill.[35] Certain later authors have credited Darwin for his methodological contribution in focusing not just on changes in appearance but also on the musculature that generated those changes, and for focusing primarily on facial expression as a reflection of underlying emotions.[36] In fact, the credit belongs to Charles Bell, whose work was published sixty-five years earlier and whose example Darwin acknowledged:

> Sir Charles Bell, so illustrious for his discoveries in physiology, published in 1806 the first edition, and in 1844 the third edition of his "Anatomy and Philosophy of Expression." He may with justice be said, not only to have laid the foundations of the subject [of expression] as a branch of science, but to have built up a noble structure. His work is in every way deeply interesting; it includes graphic descriptions of the various emotions, and is admirably illustrated. . . . The merits of Sir C. Bell's work have been undervalued or quite ignored.
>
> From reasons which will presently be assigned, Sir C. Bell did not attempt to follow out his views as far as they might have been carried. He does not try to explain why different muscles are brought into action under different emotions.[37]

As regards Darwin's last comment concerning Bell's failure to explain the reason that different muscles are activated by different emotions, the same criticism can be applied to nearly all current studies of facial expression.[38] Interestingly, Darwin believed that one of Bell's major contributions was in showing the close

---

º Duchenne, without the benefit of an official university or hospital appointment in Paris, examined the effects of electrical stimulation on normal and diseased nerves and muscles, thereby developing the techniques of electrodiagnosis and electrotherapy, and invented an instrument (resembling a harpoon) with which to perform muscle biopsies, establishing this as a new diagnostic tool. He also pioneered the medical uses of photography. He was instrumental in describing several neurological disorders, including one (now designated Duchenne muscular dystrophy) characterized by severe progressive muscle weakness in young boys, and another in which there is progressive muscle weakness and wasting due to disease of the spinal cord (Duchenne–Aran type spinal muscular atrophy).

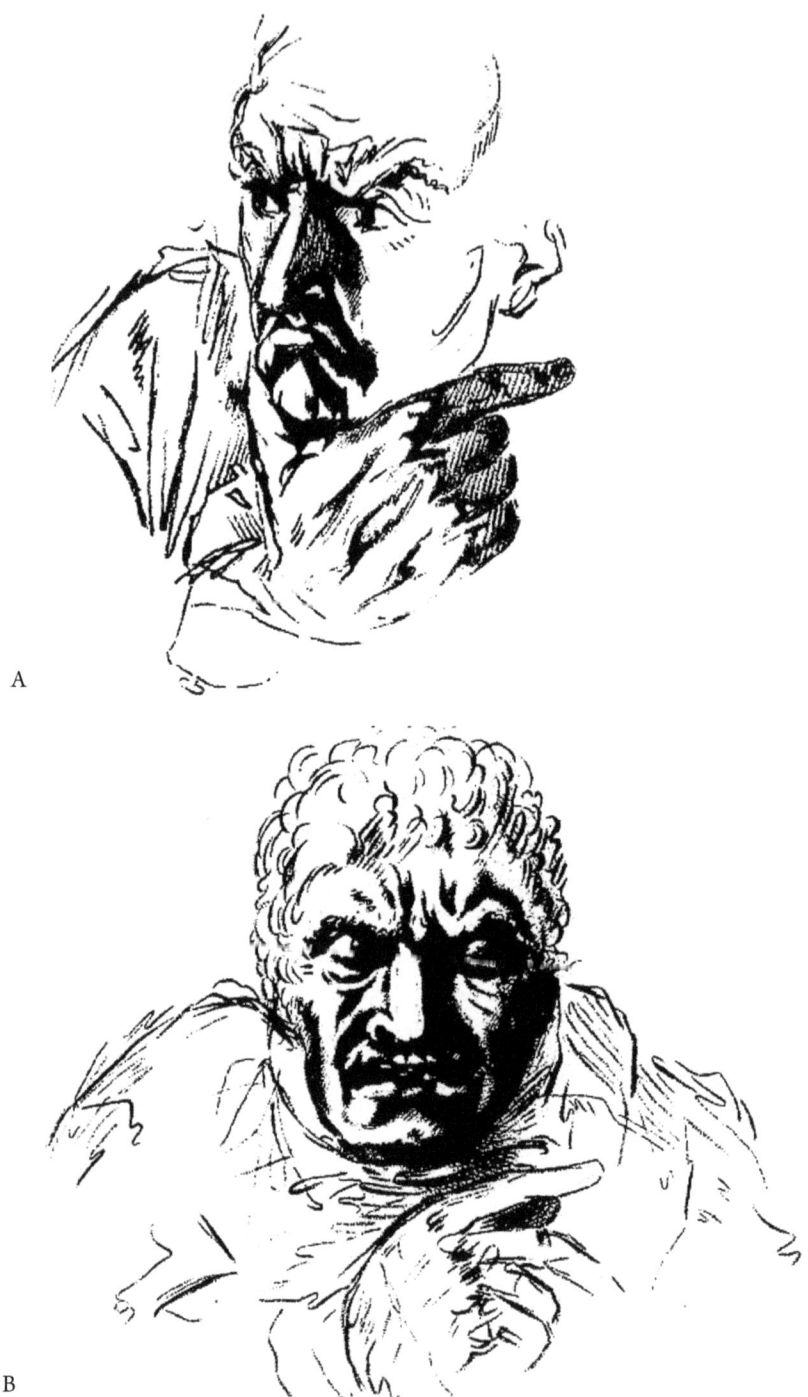

Figure 4.4 (A) Discontent. The brow is clouded, the nose peculiarly arched, and the angle of the mouth drawn down. (B) Suspicion and jealousy. Suspicion is characterized by earnest attention, with a certain obliquity of the eyes, jealousy by a more dark and frowning obliquity of the eyes, often with a cruel expression of the lower part of the face. (From Bell C: *Essays on the Anatomy of Expression in Painting.* pp. 133 and 136. Longman, Hurst, Rees, and Orme: London, 1806.)

Figure 4.5 Madness. Bell suggested that this should be conveyed by vacancy of mind and an animal passion. Muscles indicating sentiment should not be activated. Instead, there should be a vacancy to the laugh and a want of meaning to any ferociousness. The illustration shows the afflicted person in chains, as was the custom of the time. He looks disheveled, brutish, and aggressive with an animal quality to him. (From Bell C: *Essays on the Anatomy of Expression in Painting.* p. 153. Longman, Hurst, Rees, and Orme: London, 1806.) (Courtesy of the Wellcome Library, London.)

relation that exists between the movements of expression and respiration,[37] a topic discussed further in Chapter 7.

Darwin attempted to reconstruct hypothetically the evolution of various physical expressions of emotions.[39] He believed that many of these expressions are inherited, but in a manner that now seems somewhat surprising. He described three principles concerning the generation of expressions and attempted to codify the reasons that particular facial movements signify particular emotions. His notion of "serviceable-associated habits" suggested that, with repetition, a deliberate or intentional action accompanying a specific emotion becomes associated with that emotion and subsequently occurs as an expression of it. For example, humans may shut the eyes or shake the head on describing an unpleasant sight, or they may raise the eyebrows while trying to recall something, as if trying to increase the visual field.[40] His "principle of antithesis" was that some movements are performed because they are the opposite in nature to a serviceable habit but are useless in themselves (e.g., shrugging the shoulders, bending the elbows inward, raising the open hands with fingers spread, raising the eyebrows, and opening the mouth to express impotence or helplessness).[41] In both circumstances, the expressive movements were initially volitional but later, through repetition, became involuntary and heritable. This seemed to suggest the Lamarckian inheritance of acquired characteristics rather than the acquisition of behavior patterns through natural selection, a concept that was discredited within a few years of the book's publication. Darwin's rather vague third principle of "actions due to the constitution of the nervous system" referred to movements that were independent of habit or volition, such as trembling with rage, cold, or fear.[39,42] However, many psychologists have had difficulty in accepting the view that expressions are simply the inherited remnants of serviceable habits acquired during evolution.[43]

Throughout the years, there have been other concerns about Darwin's views.[38] For example, Darwin held that emotions are not unique to humans, although many biologists believed—at least until recently—that it was unacceptable to describe the actions of animals in terms of human emotion, preferring that observable behavior should be described without any inferences about motivation. Even if animals do have emotions, it is unclear how many emotions they have, how complex they are, whether they can feel their emotions, and whether they are aware of what they feel. Interestingly, Darwin may have intentionally avoided unsettling the Victorian public by arguing that humans had "animal traits" and instead charmed them by telling stories of "human traits in animals."[44]

Again, Darwin carefully examined the information conveyed with a particular expression by determining how people interpreted a photograph and whether they agreed with each other regarding the emotion being expressed. This may have been a reasonable approach in the circumstances, although a number of the photographs that he used showed staged emotions.[45] Moreover, much of the evidence was anecdotal, the number of persons studied was small, and the questions asked about the photographs were often framed to elicit a desired response.[46] Finally, Darwin rejected Bell's theological framework for understanding expressions but himself seemed not to recognize their communicative function, instead perceiving them as useless.[47] Although various physical expressions reflect the state of mind, he believed that "this result was not at first either intended or expected."[48,49] Of course, emotional expressions are indeed informative, but that does not mean that they are made with the deliberate aim to communicate with others.[50]

Darwin's book was compared adversely by some to Bell's *Essays*, which seemed to be almost a work of art in itself based on its illustrations and prose.[51] Indeed, Bell's book had cultural undertones that were lacking in that of Darwin, who simply ignored Bell's comments on art and sculpture. Some contemporary reviewers saw Darwin's book as just an extension of Bell's work, separating his evolutionary concepts from his observations on expression.[51] Others saw it as an attempt to challenge Bell's creationist views, a challenge that was deemed successful or a failure by different reviewers.[51]

It would be more than a half-century before the theme of animal expressions was revisited in detail, and another thirty years before the subject was really advanced further.[52] In any event, it is now clear that chimpanzees and certain other nonhuman primates have similar facial muscles and expressions to humans as judged by both the nature of the expressions and the social context in which they occur.[52,53] However, chimpanzees lack certain sources of contrast that in the human face may facilitate detection of movements, such as white sclera, everted lips, and hairless forehead.[52]

The psychologist Paul Ekman and his collaborators in San Francisco have utilized a similar but more formalized methodology in studying facial expressions and their relation to emotions in a variety of different cultures and found that certain basic emotions (such as anger, disgust, sadness or distress, joy, fear, surprise) are not culture specific but, rather, universal in human society.[54] Ekman applied the word "basic" to emotions that evolved for their adaptive value in dealing with fundamental life tasks such as losses, frustrations, and so on,[45,55] but he defined emotions very narrowly and excluded other emotions such as romantic or parental love or hate, envy, or jealousy as affective "commitments" or "plots."[45,55,56]

In the past forty years, the study of facial expression has moved forward, especially with recent attempts to detect deception by studying the facial expressions of emotion. Clearly, the subject has advanced since Bell's *Essays* of 1806, but despite his creationist views, Bell—by his anatomical methodology—put such studies on a sound scientific footing. In the context of the times, more than fifty years before evolution by natural selection was advanced as a scientific theory, his creationism is not unexpected and should not detract from his main thesis. Indeed, although not the first to write on facial expressions, it seems appropriate that, as suggested by Darwin, he is credited with founding the scientific study of expression.

BELL'S *ESSAYS* AND THE ARTISTIC REPRESENTATION OF EXPRESSION

Bell's book also had a long-lasting effect on British and European art. It was widely used throughout the nineteenth century by artists such as John Flaxman, Henry Fuseli, David Wilkie, Ford Madox Brown, and John Everett

Millais.[57] Benjamin Robert Haydon (see p. 32, footnote D) was an enthusiastic admirer of Bell's work, attended all of Bell's lectures, and sent his own students—including the Landseer brothers, William Bewick, and Edward Chatfield—to Bell's classes.

Bell's *Essays* also provided inspiration for the French painters Théodore Géricault (1791–1824) and Eugène Delacroix (1798–1863),[58,59] who portrayed the different emotional states of their subjects through their physical manifestations. Thus, characters in their paintings often display the mixed emotions of those facing death or the dead.[59] Gericault certainly knew Bell's book, which is said to have played a role in the evolution of *The Raft of the Medusa*, his 1819 painting that in its conception meets Bell's requirement for an artist to be able to re-create accurate images of human suffering despite any repugnance toward the unpleasant context in which they occur.[59,60] It has also been suggested that certain of the facial images portrayed by Géricault closely resemble illustrations from Bell's *Essays*, as does one of the images in *The Barque of Dante* by Delacroix,[59] although other sources of these images have also been suggested. Delacroix was a frequent visitor to the zoo, botanical gardens, and Museum of Natural History in Paris, viewing the animals (particularly the lions) and even dissecting them better to understand their anatomy and capture faithfully their expressions so as to portray them with accuracy in his paintings,[61] as did Edwin Landseer in Britain.[61]

David Wilkie—who had believed in Bell's approach since their days together in Edinburgh—was moved to suggest that paintings of history should be based on detailed observations of human behavior.[58] Some authors have suggested that Wilkie did not simply learn from Bell but was stimulated to question the borderline between art and science "within the field of visual culture."[62] Be that as it may, Bell clearly had an influence on Wilkie even if some of the faces portrayed in Wilkie's paintings fail to live up to Bell's standards.

BELL AND THE PRE-RAPHAELITES

The Pre-Raphaelites were especially influenced by Bell.[63] The Pre-Raphaelite Brotherhood (later known as the Pre-Raphaelites) consisted of a group of seven English painters, poets, and critics,[P] who came together in 1848, with a common interest in realism and also in medievalism and symbolism. They opposed Joshua Reynolds and contemporary art as it was taught by the establishment

---

[P] The seven consisted of the three founders—William Holman Hunt, John Everett Millais, and Dante Gabriel Rossetti—joined by William Michael Rossetti, James Collinson, Frederic George Stephens, and Thomas Woolner.

in the art academies, with its requirement that physiognomy and bodily proportions conform to the ideals of antiquity.[64] A loose collection of painters, sculptors, illustrators, poets, and others formed around them and became quite influential in Britain for the rest of the century. They favored a naturalistic style with bright colors, intricate detail based on a meticulous observation of nature, and—often—a religious or literary subject matter. They preferred lines and angles, reflecting motion, rather than the rounded, softer forms that characterized more classic compositions. Bell's book served to justify their approach, helping them to see and understand what might otherwise be taken for granted, and thus to paint more realistically, and enabling them to add a kinetic dimension to their art form. It encouraged the representation of fleeting or mixed expressions and provided a source book of the language of expression and movement.[65]

The realism of the Pre-Raphaelites was held by some to shock or detract from any esthetic quality, especially when much-loved figures were not idealized but portrayed warts and all. Bell encouraged the depiction of pain, suffering, disease, and the like, for he did not accept that esthetic standards required emotions to be ignored or appearance to be idealized, and the Pre-Raphaelites followed his lead, incurring much contemporary criticism.[65]

The group was encouraged and supported by the leading art critic of the period, John Ruskin (1819–1900), and their style of art remained popular for many years. The critics of Bell and the Pre-Raphaelites included the influential artist and art historian Ralph Nicholson Wornum (1812–1877), who objected strongly to applying the principles of anatomy to painting, believing that art should exemplify a world of ideal forms rather than reflecting the imperfections and mishaps of nature.[66,67] He failed to understand that anatomy dealt with more than ugliness and disease, that it involved reality in all its forms. He refused to understand that expressions and gestures are the outward reflection of an individual reality. Frank Stone (1800–1859), the art critic for the *Athenaeum*, became a bitter enemy after one of his own paintings was rubbished by a Pre-Raphaelite (Gabriel Rossetti), and he made his negative views plain when reviewing the work of Millais, Hunt, and Rossetti at the Royal Academy Exhibition of 1850. Stone was an old friend of Charles Dickens who—in the weekly *Household Words*—proceeded to excoriate the Pre-Raphaelites.[68] The family of Christ had been depicted by Millais as an ordinary family in his *Christ in the House of His Parents*. The symbolism in the painting was ignored or regarded as offensive, and Dickens complained that Mary was horribly ugly with a seemingly dislocated throat, and Jesus appeared as a blubbering boy with a wry neck wearing a bed gown. The fact that Mary—like all mothers—was leaning forward to examine a minor injury to her son and that Jesus—like all children—was bending forward for a reassuring kiss seems to have escaped the master chronicler.

Another art critic was Elizabeth Eastlake (1809–1893), who was herself married to a painter.^Q She felt distaste for the expression in art of so-called private emotions, which—unlike the intellect—did not reflect all that is noble in humanity, and certain of which were too ugly to be portrayed.[65] Thus, she too disagreed with Bell and with the painters who followed his lead by depicting expressions of emotions that she deemed less than ideal. However, her personal animosity to John Ruskin, who had criticized her husband's paintings, may have been responsible, at least in part, for her beliefs. There were also other personal issues. Ruskin was unhappily married, his marriage never consummated, his wife Effie claiming he was impotent, he claiming that she was mentally unbalanced. Effie fell in love with Millais and confided in Elizabeth Eastlake, who helped to defend her against public gossip and Ruskin. After the eventual annulment of the marriage, Millais married Effie, with whom he had several children. Ruskin himself never remarried and had several mental breakdowns before his death in 1900.

All told, Bell's influence on the visual arts was substantial, and much of the formal opposition to the Pre-Raphaelites arose from personal animosities and matters unrelated to their art. Indeed, their art is said to have influenced the arts and crafts movement that developed as a counterweight to the industrial revolution in the last decades of the nineteenth and early years of the twentieth centuries, eventually giving way to the modernist movement. Bell's own skills as an artist—and particularly his remarkable paintings of the wounds of war and their surgical treatment—are considered further in Chapter 5.

REFERENCES

1. Darlington A: The teaching of anatomy and the Royal Academy of Arts. *J Art Design Educ* 1986; 5: 263–271.
2. Ballestriero R: Anatomical models and wax Venuses: Art masterpieces or scientific craft works? *J Anat* 2010; 216: 223–234.
3. Riva A, Conti G, Solinas P, Loy F: The evolution of anatomical illustration and wax modelling in Italy from the 16th to early 19th centuries. *J Anat* 2010; 216: 209–222.
4. Bell M: Footnote. p. 73. In *Letters of Sir Charles Bell, K.H., F.R.S. L. & E: Selected from His Correspondence with His Brother George Joseph Bell*. Murray: London, 1870.
5. Bell C: *Essays on the Anatomy of Expression in Painting*. Longman, Hurst, Rees, and Orme: London, 1806.
6. Bell C: Dedication page. In *Essays on the Anatomy and Philosophy of Expression*, 2nd edition. Davison: London, 1824.

---

^Q Her husband was to become president of the Royal Academy and director of the National Gallery in London.

7. Bell C: Letter to George dated 1 January 1808. pp. 121–122. In *Letters of Sir Charles Bell, K.H., F.R.S. L. & E: Selected from His Correspondence with His Brother George Joseph Bell.* Murray: London, 1870.
8. Jordanova L: The representation of the human body: Art and medicine in the work of Charles Bell. pp. 79–94. In Allen B (ed): *Towards a Modern Art World.* Yale University Press: New Haven, CT, 1995.
9. Berkowitz C: The beauty of anatomy: Visual displays and surgical education in early-nineteenth-century London. *Bull Hist Med* 2011; 85: 248–278.
10. Bell C: Introductory essay. p. 9. In *Essays on the Anatomy of Expression in Painting.* Longman, Hurst, Rees, and Orme: London, 1806.
11. Bell C: Essay III. pp. 55–56. In *Essays on the Anatomy of Expression in Painting.* Longman, Hurst, Rees, and Orme: London, 1806.
12. Bell C: Essay V, labeled Essay VI in the original. pp. 107–157. In *Essays on the Anatomy of Expression in Painting.* Longman, Hurst, Rees, and Orme: London, 1806.
13. Loudon ISL: Sir Charles Bell and the anatomy of expression. *Br Med J* 1982; 285: 1794–1796.
14. Bell C: *Essays on the Anatomy and Philosophy of Expression,* 2nd edition. p. 141. Murray: London, 1824.
15. Gardner-Thorpe C: The art of Sir Charles Bell. pp. 99–128. In Rose FC (ed): *Neurology of the Arts: Painting, Music, Literature.* Imperial College Press: London, 2004.
16. Darlington AC: *The Royal Academy of Arts and Its Anatomical Teachings; With an Examination of the Art Anatomy Practices During the Eighteenth and Early Nineteenth Centuries in Britain.* PhD thesis, University of London, 1990.
17. Bell C: Introductory essay. p. 5. In *Essays on the Anatomy and Philosophy of Expression,* 2nd edition. Murray: London, 1824.
18. Bell C: Essay II. pp. 15–17. In *Essays on the Anatomy and Philosophy of Expression,* 2nd edition. Murray: London, 1824.
19. Bell C: Essay II. p. 21. In *Essays on the Anatomy and Philosophy of Expression,* 2nd edition. Murray: London, 1824.
20. Cummings F: Charles Bell and *The Anatomy of Expression. Art Bull* 1964; 2: 191–203.
21. Bell C: Letter to George dated 19 November 1807. pp. 114–116. In *Letters of Sir Charles Bell, K.H., F.R.S. L. & E: Selected from His Correspondence with His Brother George Joseph Bell.* Murray: London, 1870.
22. Bell C: Essay II, p. 46. In *Essays on the Anatomy of Expression in Painting.* Longman, Hurst, Rees, and Orme: London, 1806.
23. Bell C: Letter to George dated 5 December 1807. p. 119. In *Letters of Sir Charles Bell, K.H., F.R.S. L. & E: Selected from His Correspondence with His Brother George Joseph Bell.* Murray: London, 1870.
24. Taylor T (ed): *Life of Benjamin Robert Haydon, Historical Painter, From His Autobiography and Journals.* 2nd edition, Volume 1, pp. 43–44. Longman, Brown, Green, & Longmans: London, 1853.
25. Cooper A: Letter to Sir W. Beechey. p. 131. In *Letters of Sir Charles Bell, K.H., F.R.S. L. & E: Selected from His Correspondence with His Brother George Joseph Bell.* Murray: London, 1870.
26. Taylor T (ed): *Life of Benjamin Robert Haydon, Historical Painter, from His Autobiography and Journals.* 2nd edition, Volume 1, pp. 99–100. Longman, Brown, Green, & Longmans: London, 1853.

27. Bell C: Letter to George, undated ("Friday night"). p. 119. In *Letters of Sir Charles Bell, K.H., F.R.S. L. & E: Selected from His Correspondence with His Brother George Joseph Bell*. Murray: London, 1870.
28. Bell C: Letter to George dated 26 July 1808. pp. 125–127. In *Letters of Sir Charles Bell, K.H., F.R.S. L. & E: Selected from His Correspondence with His Brother George Joseph Bell*. Murray: London, 1870.
29. Lawrence T: Letter to a friend, dated 30 November 1808. Item LAW/1/202. Archives of the Royal Academy of Arts, London.
30. Carlisle A: On the connection between anatomy and the arts of design. *Artist* 1807; 17: 1–11.
31. Anon (signed "R"): Article II. A reply to "A Defence of the Royal Academy," in the 10th number of the Annals of the Fine Arts. Taken from the Times daily paper. *Ann Fine Arts* 1818; 3: 542–551.
32. Whitley WT: Turner as a lecturer. *Burlington Mag* 1912–1913; 22: 257–258.
33. Hartley L: *Physiognomy and the Meaning of Expression in Nineteenth-Century Culture*. pp. 156–157. Cambridge University Press: Cambridge, UK, 2001.
34. Darwin C: *The Expression of the Emotions in Man and Animals*. Murray: London, 1872.
35. Black J: Darwin in the world of emotions. *J R Soc Med* 2002; 95: 311–313.
36. Ekman P: Darwin's contributions to our understanding of emotional expressions. *Philos Trans R Soc Lond B* 2009; 364: 3449–3451.
37. Darwin C: Introduction. pp. 2–3. In *The Expression of the Emotions in Man and Animals*. Murray: London, 1872.
38. Ekman P: Introduction to the third edition of *The Expression of the Emotions in Man and Animals* (by Charles Darwin), anniversary edition. pp. xxi–xxxvi. Oxford University Press: New York, 2009.
39. Browne J: Darwin and the expressions of the emotions. pp. 307–326. In Kohn D (ed): *The Darwinian Heritage*. Princeton University Press: Princeton, NJ, 1985.
40. Darwin C: *The Expression of the Emotions in Man and Animals*. pp. 32–33. Murray: London, 1872.
41. Darwin C: *The Expression of the Emotions in Man and Animals*. pp. 264–265. Murray: London, 1872.
42. Darwin C: *The Expression of the Emotions in Man and Animals*. pp. 67–68. Murray: London, 1872.
43. Carmichael L: Sir Charles Bell: A contribution to the history of physiological psychology. *Psychol Rev* 1926; 33: 188–217.
44. Korn E: How far down the dusky bosom? *Lond Rev Books* 1998; 20(23): 23–24.
45. Gross DM: Defending the humanities with Charles Darwin's *The Expression of the Emotions in Man and Animals* (1872). *Critical Inquiry* 2010; 37: 34–59.
46. Ekman P: Introduction, afterword, and commentaries on *The Expression of the Emotions in Man and Animals* (by Charles Darwin), anniversary edition, pp. 366–367. Oxford University Press: New York, 2009.
47. Hartley L: *Physiognomy and the Meaning of Expression in Nineteenth-Century Culture*. p. 146. Cambridge University Press: Cambridge, UK, 2001.
48. Darwin C: *The Expression of the Emotions in Man and Animals*. p. 357. Murray: London, 1872.
49. Burkhardt RW Jr: Darwin on animal behavior and evolution. pp. 327–365. In Kohn D (ed): *The Darwinian Heritage*. Princeton University Press: Princeton, NJ, 1985.

50. Ekman P: Introduction, afterword, and commentaries on *The Expression of the Emotions in Man and Animals* (by Charles Darwin), anniversary edition, p. 373. Oxford University Press: New York, 2009.
51. Smith J: *Charles Darwin and Victorian Visual Culture*. pp. 186–198. Cambridge University Press: Cambridge, UK, 2006.
52. Vick S-J, Waller BM, Parr LA, Smith Pasqualini MC, Bard KA: A cross-species comparison of facial morphology and movement in humans and chimpanzees using the Facial Action Coding System (FACS). *J Nonverbal Behav* 2007; 31: 1–20.
53. Waal FBM. Darwin's legacy and the study of primate visual communication. *Ann NY Acad Sci* 2003; 1000: 7–31.
54. Ekman P, Oster H: Facial expressions of emotion. *Annu Rev Psychol* 1979; 30: 527–554.
55. Ekman P: Basic emotions. pp. 45–60. In Dalgleish T, Power M (eds): *Handbook of Cognition and Emotion*. Wiley: Chichester, UK, 1999.
56. Ekman P: Introduction, afterword, and commentaries on *The Expression of the Emotions in Man and Animals* (by Charles Darwin), anniversary edition. Commentary, pp. 83–84. Oxford University Press: New York, 2009.
57. Codell JF: Empiricism, naturalism and science in Millais's paintings. pp. 119–147. In Mancoff DN (ed): *John Everett Millais: Beyond the Pre-Raphaelite Brotherhood*. Yale University Press: New Haven, CT, 2001.
58. Dawson D, Morère P: Introduction: A new incarnation of the auld alliance: "Franco-Scottish studies." p. 21. In Dawson D, Morère P (eds): *Scotland and France in the Enlightenment*. Associated University Presses: London, 2004.
59. Macmillan D: French art and the Scottish enlightenment. pp. 128–158. In Dawson D, Morère P (eds): *Scotland and France in the Enlightenment*. Associated University Presses: London, 2004.
60. Moscoso J: The shadows of ourselves: Resentment, monomania and modernity. pp. 23–28. In Fantini B, Moruno DM, Moscoso J (eds): *On Resentment: Past and Present*. Cambridge Scholars: Newcastle upon Tyne, UK, 2013.
61. Jackson D: *Lion*. pp. 85–88. Reaktion: London, 2010.
62. Tromans N: The anatomy of expression. pp. 61–113. In *David Wilkie: The People's Painter*. University of Edinburgh Press: Edinburgh, 2007.
63. Hartley L: *Physiognomy and the Meaning of Expression in Nineteenth-Century Culture*. p. 79. Cambridge University Press: Cambridge, UK, 2001.
64. Bullen JB: *The Pre-Raphaelite Body: Fear and Desire in Painting, Poetry, and Criticism*. pp. 9–10. Clarendon: Oxford, 1998.
65. Codell JF: Expression over beauty: Facial expression, body language, and circumstantiality in the paintings of the Pre-Raphaelite Brotherhood. *Victorian Studies* 1986; 29: 255–290.
66. Wornum RN: Modern moves in art. "Christian architecture." "Young England." *Art-Journal* 1850; 12: 269–271.
67. Hartley L: *Physiognomy and the Meaning of Expression in Nineteenth-Century Culture*. p. 91. Cambridge University Press: Cambridge, UK, 2001.
68. Dickens C: Old lamps for new ones. *Household Words* 1850; 12 (June 15, 1850): 12–14.

5

BEHIND THE GLORIES OF WAR

Bell's accomplishments as an artist were not simply academic. He established himself as a talented sketcher whose work is characterized by the realism of expression and form that he wrote about in his *Essays*. His paintings and drawings reflect his belief that the visual arts should portray the imperfections and mishaps of life rather than an idealized world of make-believe. His better known artistic works were made against the backdrop of a destructive war that continued in Europe for years and reshaped the continent. They stand in bitter contrast to the grand and inspiring martial scenes usually displayed, reflecting the bleak horror behind the glories of war. They can be appreciated only by understanding the context in which they were made. Appendix 3 lists Bell's paintings, which can be viewed in museums as well as online in digitized form.

THE RETREAT TO CORUNNA

The Iberian Peninsula became a major theater of war between Britain and Napoleonic France during the first decade of the nineteenth century, and Bell came to be involved especially in caring for the wounded of Corunna. Indeed, some of his most important artistic compositions relate specifically to his experiences in caring for the wounded from that battle and, later, at Waterloo.

The Peninsular War began when allied French and Spanish forces invaded Portugal—Britain's oldest ally in Europe—in November 1807 and continued until Napoleon's defeat in 1814. Corunna—a port on the northwestern Spanish coast—was the site of a rearguard action that was both a disaster and a victory for the British under Lieutenant-General Sir John Moore,[A] commander of British

---

[A] Lieutenant-General Sir John Moore (1761–1809) joined the army in 1776, serving in America during the War of Independence. He served as member of parliament for Lanark Burghs (also known as Linlithgow Burghs, in Scotland) between 1784 and 1790. He subsequently served with the army in Corsica, the West Indies, Ireland, The Netherlands, Egypt, and the Mediterranean. In 1801, he was appointed colonel of the 52nd Foot regiment. On the renewal of war with France in 1803, he was appointed to a brigade in Kent, England, and introduced an innovative training regime that led to Britain's first permanent light infantry regiments. Knighted in 1804, he was given field command in the Iberian Peninsula and was killed in 1809 at the Battle of Corunna.

forces on the Iberian Peninsula. In 1808, France turned on Spain, its former ally, and Moore's army moved from Portugal to the support of Spanish forces fighting against Napoleon's invading armies. With the Spanish overwhelmed and in disarray, the grossly outnumbered and weary British—cut off from their own supply lines—were forced to retreat or face huge losses. Their line of retreat into Portugal was already cut, so they made their way in dreadful weather and with great hardship over 250 miles across the mountains to Corunna, despite constant harassment by the French. The sick and injured were left to their fate by the wayside. Discipline collapsed; drunkenness, looting, and desertion became an increasing problem. The disintegrating army, which had already lost several thousand men, finally arrived at the port city looking so pitiful that the townspeople made the sign of the cross as the troops passed.[1] They arrived ahead of the fleet that was to return them to England, allowing the French to catch up and then—on 16 January 1809—attack them.

The British counterattacked and were able to drive back their enemy. Much of the army (approximately 80 percent) eventually embarked—under cover of heavy naval guns and supported by their Spanish allies—on the transports that had arrived with the fleet's tardy appearance. The luckless Moore, however, took a cannonball in the left shoulder; he was carried back to his quarters in the city, where he was laid on a mattress and died after several hours, having lived to hear that the French were beaten and in full retreat. He was buried in Spain, wrapped in a military cloak and blankets. A famous poem, known to generations of British schoolchildren, recalls the event.[B] The army sailed unmolested for England that same day.

Although the casualties of the French under Marshal Soult[C] were considerably heavier than the British, the British had been driven from Spain at least for the time being. As a sign of respect to his opposite number, Soult had a monument built to Moore. But Moore's reception back home was more ambiguous. The

---

[B] The poem, by the Irishman, Charles Wolfe, is titled *The Burial of Sir John Moore After Corunna*. The first stanza reads

> Not a drum was heard, not a funeral note,
> As his corse to the rampart we hurried;
> Not a soldier discharged his farewell shot
> O'er the grave where our hero we buried.

[C] Marshal General Jean-de-Dieu Soult (1769–1851), French soldier and statesman, served Napoleon and then Louis Philippe, whom he represented at the coronation of Queen Victoria. He was very popular with the British because, after the battle of Corunna, he instructed his artillery to fire a salute to their bravery and erected a memorial to Moore. It is said that at a party during the coronation celebrations, he came face to face with Wellington, his former opponent, who exclaimed "I have you at last!"

exhausted troops that arrived back on the south coast of England—some twenty-eight thousand—were filthy, dejected, and disorderly, and perhaps some six thousand were sick or injured, their discomfort aggravated by the four or five days of the overcrowded and stormy voyage home, during which the sick were mixed with the healthy, the living with the dead or dying. Argument and recriminations about the defeat and behavior of the troops soon followed. Some held that Moore had mishandled the retreat, but most people held him blameless, and—with the discovery of a copy of his diaries many years later—his name and achievements have subsequently come to be honored.

Sir Arthur Wellesley (1769–1852; later the Duke of Wellington), who had previously served in Portugal, returned there and to the Peninsula campaign in 1809. Both the war, which dragged on until the defeat of Napoleon in 1814, and its aftermath had their own horrors. The effects on the civilian population—physical suffering and humiliation—were depicted by Francisco Goya (1746–1828) in a series of etchings and aquatint prints (*The Disasters of War*) that were not released until after his death. The horrors for the soldiers were just as bad as can be gleaned from some of Bell's paintings. The poor level of the medical services of the time was compounded by inadequate supplies, equipment, and transport. The nature of sepsis and of infections was poorly understood, anesthesia was not available, and nursing care was rudimentary. The training of military surgeons was haphazard and without any set curriculum, involving an apprenticeship followed by an oral examination. Training was the same for regimental or staff surgeons, but the former then were exposed to only a limited variety of cases, whereas the latter gained more general experience.[2]

MEDICAL CONDITIONS AT THE TIME OF CORUNNA

The medical department of the British Army at the time of Corunna was administered by a board consisting of the physician-general (Sir Lucas Pepys[D]), surgeon-general (Thomas Keate[E]), and inspector-general of army hospitals (Francis Knight). The physician-general appointed all the physicians in the army, and the surgeon-general appointed all the surgeons; both were civilians. The board was incompetent, and its members quarreled among themselves and had little understanding of military matters. It was responsible for the provision of medical care and maintaining professional medical standards in the army, but

---

[D] Sir Lucas Pepys (1742–1830), physician to the Middlesex Hospital and president of the Royal College of Physicians of London, attended King George III for his mental disorder and appeared before a committee of the House of Commons to discuss the king's illness. In 1794, he became physician-general to the army and president of the Army Medical Board.
[E] Thomas Keate (1745–1821), surgeon to the Prince of Wales, afterwards King George IV, succeeded John Hunter in 1793 as surgeon-general to the army.

especially for the administration of general hospitals and the medical services of large units on campaign. In the field, medical officers were divided into staff officers working in general hospitals or as administrators under the direction of the inspector-general and regimental officers who treated members of their unit in their own hospitals and were responsible directly to their regimental commander. The physicians generally possessed good academic (but few military) qualifications, were well paid (one pound daily), and personally attended the field commander or supervised large hospitals.[F] The staff surgeons, who did not belong to a particular regiment, were aided by hospital mates (junior staff, later called hospital assistants) from among whom they were often selected. The regimental surgeons were required to manage expenditures and to record patients and their progress in a register that included all medications and daily diets. Regimental hospitals were not mobile, so the sick or injured had to be treated at general hospitals set up in towns, military bases, or along lines of communication when the army was on the march.[1]

In fact, the sick and wounded on the peninsula were often housed in any available building—convents, barracks, farms, private houses, or barns—or in the carts and wagons in which they had been transported, and even the most basic amenities, such as water, soap, chamber pots, towels, or bedding, were lacking. Nevertheless, the "gentlemen of the medical department in Spain" did what they could under difficult circumstances and with "great zeal and humanity."[3] Indeed, it has been said that more soldiers died from administrative incompetence than from lack of surgical expertise. Back in England, a shortage of hospital beds also became acute with the return of the army from Corunna, especially as Francis Knight, the inspector-general, had previously closed several general hospitals in an economy drive, leading to a general shortage of beds.

Most of the returning sick and injured were landed at Portsmouth or Plymouth after delays caused by bad weather, during which healthy and sick were mixed so that the spread of disease was actually encouraged. Many died in the boats that took them to shore and were brought as corpses or in a preterminal state to the hospitals. The deputy inspector of hospitals at Portsmouth, James McGrigor,[G] laid

---

[F] A good account of the ranks and appointment process in the army before and during the Peninsula campaign is provided by Howard M: *Wellington's Doctors: The British Army Medical Services in the Napoleonic Wars*. Spellmount: Staplehurst, UK, 2002.

[G] Sir James McGrigor (1771–1858) was a Scottish military surgeon who was deputy inspector-general (subsequently promoted to inspector-general) of hospitals at Portsmouth during the evacuation of the army from Corunna, later took charge (as chief inspector of hospitals) of the medical department of Lord Wellington's army during the Peninsular War, and—a brilliant administrator—eventually became director general of the army's medical department, a post he held until 1851. Further details are provided by Blanco RL: *Wellington's Surgeon General: Sir James McGrigor*. Duke University Press: Durham, NC, 1974.

on extra personnel by enrolling officers from the Household Troops and taking on local practitioners. Despite using barracks for the overflow from hospitals and also utilizing the Haslar Hospital (in nearby Gosport), the facilities were soon completely overwhelmed, and transport and prison ships were pressed into service. Knight even requested London medical students to come to the aid of the sick. More soldiers were sick than wounded—fever, dysentery, cholera, and typhus turned whispers of mortality into a death rattle. Richard Hooper, a surgeon of the time, described patients with dysentery:

> The stools were very thin, frequent . . . tinged with black mucus, and accompanied by a most painful tenesmus: the countenances much emaciated and dejected; their bodies and clothes dirty in the extreme, and abounding with vermin; much enfeebled, and complaining of severe pains across the loins, and in the calves of the legs; considerable thirst; little or no appetite.[4]

Their treatment was with purgatives (such as calomel) and emetics, followed by pulvis ipecacuanhae compositus (compound powder of ipecacuanha), which was subsequently replaced by opium pills with greater—but still only temporary—benefit. Tincture of opium, rectal starch injections, rectal or oral "acetite of lead," wine, and decoction of bark were sometimes used, but with limited effect. The disease remitted and then recurred, and all the affected patients were eventually collected into one hospital to await their fate.

Typhus—with its fever, delirium, pain, headache, and skin rash—was common, as was pneumonia and fever of unclear origin. Gangrene in the feet and legs was also a frequent problem, and necrotic abscesses were common on other parts of the body. The mortality was high. Those who recovered were sent to a convalescent hospital and then on to different barracks where they remained until they could rejoin their regiments. But relapses in fever and dysentery occurred frequently and often led to pneumonia. Treatment varied with the physician—cordials (a mixture of water with wine or spirits) and stimulants at the Haslar Hospital (where Bell served), cold affusion at the general hospital:

> I recollect one case in particular, where, on admission, the pulse was hardly perceptible; the extremities were cold, and the patient appeared to be rapidly sinking. After the warm-bath, he was put to bed, and, as deglutition [swallowing] was easily performed, he had stimulants freely given; the pulse soon rose; the heat of the surface became intense. In this state, the cold affusion was instantly used, and, with the happiest effect.[3]

Cold affusion typically involved dashing ice-cold water violently from a sponge over the face, head, chest, and body, and then over the back, following which the patient was rubbed dry vigorously and the process then repeated two or three times.

McGrigor went on to improve the medical services in the field over the following three years or so—wards were to be fumigated with nitric acid, plaster walls whitewashed, and wooden pallets cleaned with soap and water. They were to be better ventilated and overcrowding avoided, with patients each being allocated at least 5 feet of space. Furthermore, no animals were to be kept within hospitals, codes of behavior were developed, and consumption of alcohol was forbidden.

BELL AND THE WOUNDED OF CORUNNA

On 22 January 1809, Bell wrote to his brother George that he was on his way from London to Portsmouth through heavy snow drifts to help with the wounded just back from Corunna. He had been to the Medical Board and was offered charge of the Haslar Hospital in Gosport, but he turned it down because he had to be free to "return to my lectures."[5] The Royal Hospital Haslar, a four-storey red-brick building with sash windows, set in large grounds, had been built on agricultural land more than fifty years earlier to provide improved medical care for naval personnel. Once the leading military hospital in the land, it was intended originally to accommodate fifteen hundred patients but soon housed many more plus the staff and their families.[H] James Lind was an early chief physician there and served for 25 years, to be succeeded by his own son.[I] The patients, many of whom had been press-ganged into service, might have been tempted to slip away as they recovered, so the ground-floor doors of the hospital were locked at night and soldiers patrolled the walls to prevent desertion.[6] The patients were brought in by handcarts, evaluated by a hospital mate, and assigned a place to stay. The dead and dying were also brought in, and the grounds served in part as an unconsecrated graveyard for those for whom help was ineffectual or too late.

On 3 February 1809, Bell wrote to George from Haslar Hospital, upset by what he heard and saw, overwhelmed perhaps by the sheer numbers of patients requiring medical or surgical treatment, and angered by the lack of facilities for immediate care:

> I wish I had written to you during my first sensations—these were, I trust, such as every good man should feel; they are blunted by repetition, . . . I have muttered bitter curses

---

[H] The Royal Hospital Haslar was closed in 2009.
[I] James Lind (1716–1794), Scottish surgeon, served as chief physician to the Royal Hospital at Haslar for 25 years. He established the value of citrus fruits in preventing scurvy using the methodology of a controlled clinical trial, a new approach to clinical research.

and lamentations, have been delighted with the heroism and prowess of my countrymen, and shed tears of pity in the course of a few minutes. I find myself, my dear George, in a situation unexpected and strange, such as I hope you may never see. I have stooped over hundreds of wretches in the most striking variety of woe and misery, picking out the wounded. Each day as I awake, still I see the long line of sick and lame slowly moving from the beach: It seems to have no end.[7]

A few days later, he wrote from London with anger about the way the men had been led and provided for during the retreat, echoing a common feeling but one that later proved unjustified; he also exclaimed about the disorganized rabble they had become:

I know nothing of Portsmouth. I was only one hour out of the hospital, and came away at eleven o'clock at night. I know nothing but what the world is well acquainted with— that all our soldiers are heroes, and that our generals are fools. I speak undoubtedly from the impression of a dispirited and harassed army; yet there is a show of sense that looks like truth. . . . The men assure me they marched five times over the same ground. In their retreat they had nothing provided for them. After starving for two days they got rum to comfort them; no sooner were they halted to eat than they were forced off the ground. . . . The disorders of the soldiers were shocking. I found a fellow with a stocking full of dollars, and another with a silver spoon under his pillow; the rogue grinned and said he took it from a Spaniard. . . . I have got some noble specimens of injured bones, and a series of cases admirably fitted for lectures, better fitted for my Surgery, I mean my Second Edition.[8]

By May 18, he was writing to George about a new artistic project based on his experiences with the wounded of Corunna, but also a need for funds:

I have etched several of my plates of gunshot wounds. They will be very fine, I think. I have got into the use of the needle. It is better to make these sketches good than to publish on the Anatomy of Painting. Will you pay for some of my proofs coming down?[9]

Within a week, his plans had become clearer, as he wrote on May 27:

I intended to make an appendix of gun-shot wounds, and in my second edition to throw that part of my Surgery and the cases together into a complete treatise; but the booksellers wish it to be published separately . . . so you will find me the author of a book of very interesting cases, with, I expect, fine etchings.[10]

One wonders whether it was Bell or his publishers who wished for a separate book on gunshot wounds. In any event, Bell was now spending much time sketching and caring for the injured and maimed from the war in Europe at the York Hospital, a military hospital and training center created from the Star and Garter Tavern and situated on what was then known as Five Fields and is now part of Eaton Square in London.

Between his clinical work and writing, Bell had little time for relaxation, but one of his diversions involved the dissection of a lioness, which he then made the subject of a lecture and an oil painting. He also made paintings from his sketches of gunshot wounds. He was living a "monkish life" but was not unaware of other pleasures. As he wrote in a letter to his brother on 23 May 1809,

> All that I have yet seen of English ladies—only obscurely—still serves to confirm me in the belief that they are perfection in form and mildness of spirit; something so calm and serene, so soft and ladylike, yet so dignified . . . I must not go on at this rate, else the wounded at Corunna may go hang.[11]

His expertise in etching improved, and he bought etchings by the best old masters to encourage himself. Looking back at some of his own first plates, he saw their inadequacies and rejected them. He remained poor, and his poverty was sometimes distressing, "but only on certain occasions when my character outruns my means," as he put it to George. Later in 1809, he again wrote to George indicating that he was "living very economically—of necessity indeed, being so much occupied."

After his service to the Corunna wounded and based on a number of other patients he subsequently saw in his practice, Bell prepared a *Dissertation on Gun-Shot Wounds*, which was published in 1814 as an appendix to his *System of Operative Surgery* and also as a stand-alone piece, with a number of illustrative plates (which now hang on the walls of the Royal College of Surgeons of Edinburgh). He attempted to define principles for treating these injuries but emphasized the need of a good knowledge of the anatomy of the affected parts for treatment to be effective. Different sorts of wound, affecting different parts of the body under a variety of circumstances, were considered. He also stressed the importance of sometimes doing nothing:

> We have to combat a natural desire in the attendants of doing something, and an expectation on their parts that the surgeon has an operation to perform. It requires address to compose the patient, and to convince the friends that nothing ought to be done.[12]

The work was well received, attracting favorable notice from reviewers:

> We are happy we have met with a book in which we have found a great subject opened, and questions of high importance relative to military surgery discussed.

We could wish that the author whose mind seems to be full of views on these subjects which appear to us to be correct and judicious, should extend his observations, and in the same spirit of good sense, which has dictated the remarks we have just perused, should treat the matter more in detail.[13]

Bell sketched some of his patients and their wounds, and he subsequently made oil paintings of them (listed in Appendix 3), which he annotated. These serve as a graphic record that was a valuable teaching aid in the past and remains a remarkable historical testament to the wounded of the Peninsular War and their contemporary surgical management. Several of these illustrations were used by Bell in his textbooks and monographs, in which case descriptions are also provided. These demonstrate Bell's clinical expertise. For example, one of the paintings is of a patient with a gunshot wound to the humerus. In his accompanying note, Bell notes that a finger can be introduced into the seemingly trifling wound to determine whether the bone is shattered, in which case jagged fragments will be felt and amputation will be required; if not, conservative treatment rather than amputation may be sufficient. Finally, the sketches and paintings illustrate his artistic skill in portraying the wounded with the brutal realism that he had encouraged in his book on painting. His subjects—gaunt and mutilated—lie with faces contorted by pain, eyes sunken by despair, ashen and indifferent as death approaches. Their shattered limbs and infected, necrotic wounds are shown against a stark background so that nothing can distract from the horror of what is portrayed.

THE BATTLE OF WATERLOO

The Napoleonic Wars did not come to a close until 1815. Exiled to the island of Elba after his defeat and abdication in April 1814, Napoleon escaped in the following February and returned to Paris to lead new armies against a coalition of forces in his so-called Hundred Days. This new coalition—and in particular a multinational force (British, German, and Dutch) under Wellington and a Prussian army under Marshal Blücher—decisively defeated him at the battle of Waterloo on June 18, but the outcome of the battle was close. The losses on both sides were horrific. Wellington is said to have lost twenty-nine percent of his army.[14] The following day, his army and some of its medical personnel moved on toward Paris. For the surgeons who remained behind to work on the wounded, the task seemed endless. There were no proper transports to evacuate the injured to hospitals.[j] Every building in every hamlet and village was filled with casualties,

---

[j] The French, by contrast, had developed an efficient system by which the wounded were taken away from the battle front by stretcher parties to horse-drawn ambulance carts.

and aproned surgeons in their shirtsleeves operated under makeshift conditions without a break, hastily amputating injured limbs in homes, chateaux, farms, stables, sheds, barns, churches and convents, and even in courtyards. The local inn, which had been Wellington's headquarters, became a field hospital, as nearly all the staff were wounded.[15] Sources of water within three miles of the battlefield were said to be polluted with corpses.[16] The local peasants were gathered together and made to collect and burn or bury the putrefying bodies of men and beasts.[15] Days after the fighting, wounded survivors from Napoleon's army, scattered over the countryside, were still being rounded up and taken to Brussels, Antwerp, and neighboring towns, where the inhabitants generously gave them and the allied injured clothing, sustenance, and accommodation; indeed, the doors of their houses were labeled with the number of their contained wounded.[17]

Bell heard of the battle on June 22, and four days later—reluctant to miss the chance of seeing more gunshot wounds—traveled to Brussels with John Shaw, his brother-in-law, to volunteer his services. In their haste, passports were forgotten, but their surgical instruments allowed them to pass freely. In Brussels, the wounded were everywhere. Bell spent his time there in "four great hospitals," and the stress, sights, and new experiences left him sleepless. In his notebook, he describes a wounded officer pinned to his horse by a sword that pierced the back and upper part of his thigh, passed through the leather and woodwork of his saddle, and entered the horse's body. But he was shocked that even eleven days after the battle, arrangements were still being made for the reception of the wounded. On July 3, he arose at 4 o'clock in the morning and wrote to the surgeon-in-chief, offering to perform the necessary operations on the wounded French. As he later wrote from Soho Square to Francis Horner,[K] a member of parliament,

> I found that the best cases, that is, the most horrid wounds, left totally without assistance, were to be found in the hospital of the French wounded; this hospital was only forming. They were even then [five days after his arrival] bringing in these poor creatures from the woods. It is impossible to convey to you the picture of human misery continually before my eyes. What was heart-rending in the day was intolerable at night; . . .
>
> All the decencies of performing surgical operations were soon neglected. While I amputated one man's thigh, there lay at one time thirteen, all beseeching to be taken next; one full of entreaty, one calling upon me to remember my promise to take him,

---

[K] Francis Horner (1778–1817), Scottish lawyer, Whig politician, influential political economist, and one of the founders of the Edinburgh Review. A marble statue of him by Sir Francis Chantrey stands in Westminster Abbey.

another execrating. It was a strange thing to feel my clothes stiff with blood, and my arms powerless with the exertion of using the knife! . . .

But there must ever be associated with the honours of Waterloo, to my eyes, the most shocking sights of woe, to my ear accents of entreaty, outcry from the manly breast, interrupted forcible expressions of the dying, and noisesome smells.[18]

He made careful surgical notes of the cases he encountered, accompanied by sketches of the wounded that he afterwards reproduced as watercolors (Figs. 5.1–5.3). In a separate notebook he kept a diary of his activities. He found the wounded French a strong and hardy bunch, brave and unsubdued, which looked "capable of marching, unopposed, from the west of Europe to the east of Asia." This grudging admiration was tempered by a detestation of the enemy as a "race of trained banditti, " as he wrote to George.[19] His letter was quite detailed, and George passed it on to his friend, the writer (Sir) Walter Scott, who became quite excited by it and set out some days later for Brussels and Paris, his first trip overseas. Scott recorded his experiences (including a

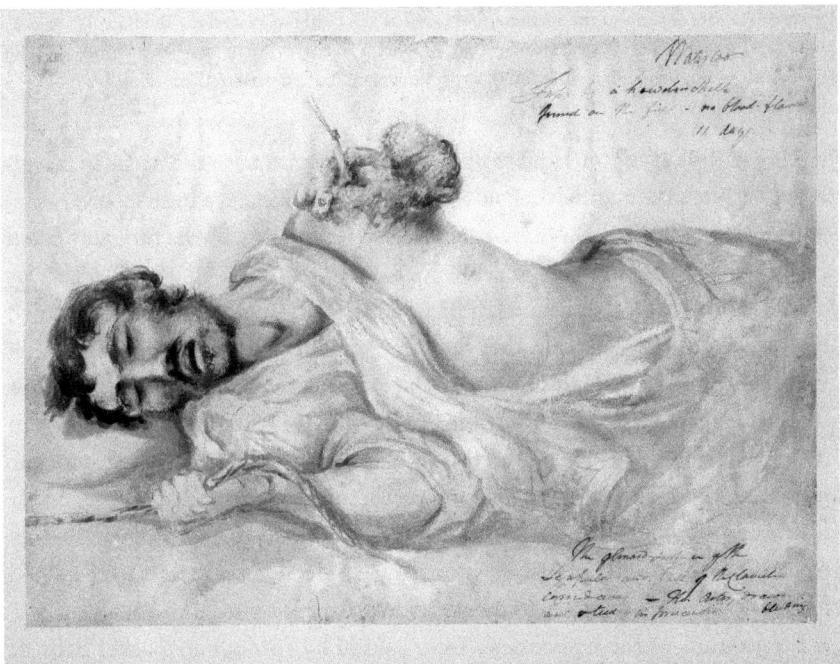

Figure 5.1 Cannon-shot wound to arm of a soldier at the battle of Waterloo. The left arm is missing and part of the left shoulder has also been carried off by cannon shot. A ligature has been tied around the left axillary artery to prevent hemorrhage. The soldier is lying on his side, grasping a rope to help him maneuver. He eventually did well. Watercolor by Bell. (Courtesy of the Trustees of the Army Medical Services Museum; Royal Army Medical Corps Muniment Collection, Wellcome Images.)

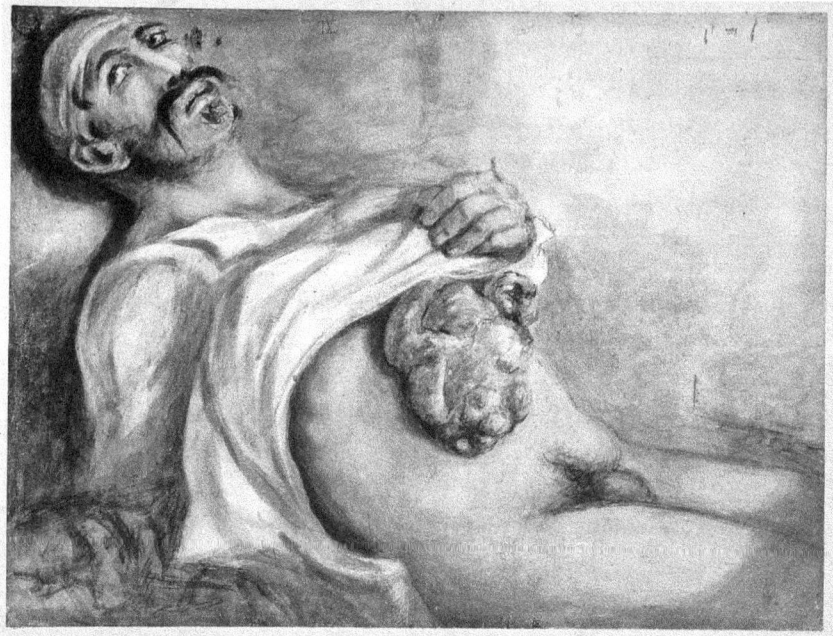

Figure 5.2 Sabre wound to the abdomen received at the battle of Waterloo. The bowels are protruded and gangrenous. Watercolor by Bell. (Courtesy of the Trustees of the Army Medical Services Museum; Royal Army Medical Corps Muniment Collection, Wellcome Images.)

meeting with Wellington) and the historical background to the battle in imaginary letters he put together as *Paul's Letters to His Kinsfolk*, a best-seller.[L]

The role that John Shaw played at Waterloo is not clear, but he probably acted as Bell's surgical assistant.[20] Curiously, Robert Knox (1791–1862), who was another of his surgical assistants at Brussels, later claimed that Bell performed only twelve amputations and that only one patient survived,[21] an observation that seems at odds with all other available descriptions. Pettigrew in 1839, for example, reported that he "afforded his professional assistance to no fewer than 300 men."[22] Because of his late arrival in Belgium, however, he was responsible for treating the French, many of whom were injured particularly gravely and for most of whom medical aid had been delayed for days, worsening their already grim outlook. It would thus not be surprising if his operative results were indeed poor and associated with a high mortality.

The wounds of the combatants were caused by cannon or musket balls or grapeshot, or by sabres, lances, and bayonets, and were complicated by infection

---

[L] In his *Memoirs of the Life of Sir Walter Scott, Bart* (Carey, Lea, and Blanchard: Philadelphia, 1837, p. 573), J. G. Lockhart points out that "parts of the great surgeon's simple phraseology are reproduced, almost verbatim, in the first of *Paul's Letters to his Kinsfolk*".

Figure 5.3 Soldier with sabre wound to the head. A portion of the skull at the vertex is completely detached by the sabre cut, but the scalp remains connected by a small isthmus. There are also facial injuries. Watercolor by Bell. (Courtesy of the Trustees of the Army Medical Services Museum; Royal Army Medical Corps Muniment Collection, Wellcome Images.)

and gangrene. Treatment involved the removal of missiles, amputation, ligation of blood vessels, and wound dressings. In many cases, amputation was probably unnecessary but resorted to speedily by inexperienced military surgeons. The stoicism of many of the patients is hard to believe. After the amputation of his arm because of a shattered elbow, the future Lord Raglan—silent until then—is said casually to have called out, "Hello, don't carry away that arm until I have taken off the ring."[15] The worst surgical cases among the British wounded were sent to the York Hospital in Chelsea, where two of the wards were under the charge of George James Guthrie, a military surgeon of distinction.

Bell was not a military surgeon but a civilian volunteer whose presence among the casualties of war from Corunna and at Waterloo was given especial notoriety because of his eminence. He had little firsthand experience of dealing with war wounds, however, and like many such volunteers, his clinical judgment and management of cases may not always have been ideal. He and the other medical volunteers surely helped to save many lives, but it seems likely that some resorted to amputation and other radical surgical measures rather too easily and perhaps unnecessarily. Larrey and Percy among the French and Guthrie and Hennen among the British probably best typify the great military surgeons of the period.

Baron Dominique Jean Larrey (1766–1842) was at one time chief surgeon of Napoleon's Grand Army, and it was he who introduced field ambulances and hospitals to the battlefield (whereas previously they were about three miles behind the line). He can be said to have originated the concept of first aid by removal of the wounded from the battlefield immediately rather than after the battle was over. He was wounded and captured by the Prussians during the retreat from Waterloo and was to have been executed, but it transpired that during the Austrian campaign he had saved the life of the son of Marshal Blücher, their commander. The Baron was released, breakfasted with the Marshal, and was escorted back to France.[M] Pierre-François Percy (1754–1825) also served for a time as chief surgeon of the Grand Army and designed a wagon to transport medical staff and their supplies to the front so wounds could be treated immediately (an approach that was less successful than that of Larrey). The names of both Larrey and Percy are inscribed on the Arc de Triomphe.

As for George James Guthrie (1785–1856), he served as an army surgeon in the Peninsular campaign where, on one occasion, he had on his hands three thousand wounded, and pioneered new surgical techniques especially for amputation, including at the hip and shoulder. At the end of the campaign, he entered private practice in London and attended Charles Bell's lectures at Great Windmill Street. After Waterloo, he went to Brussels to help the wounded and then returned to London to take charge of two wards at the York Hospital, where the most difficult cases were sent to him and where he taught for two years. He founded the Royal Westminster Eye Hospital and in 1823 was appointed as surgeon to the Westminster Hospital. He was four times president of the Royal College of Surgeons. It is said that he refused the offer of a knighthood by the Duke of York in 1826 because he was too poor.[23,24] His colleague and another illustrious military surgeon, John Hennen, graduated in Edinburgh and served through the Peninsular War in various regiments and at Waterloo. His *Practice of Military Surgery* (first published in 1818) became a standard medical work. He was known for always having a cigar in his mouth, even when operating—a habit that, surprisingly, was tolerated even by Wellington.

[M] Napoleon bequeathed in his will a large sum of money to Larrey with the comment that "he is the worthiest man I ever met." Some years later, when Napoleon's body was returned to France, Larrey, in the uniform of the Imperial Guard that he had worn at the battle of Wagram, followed the remains to its tomb.

Bell's dissertation on gunshot wounds served as a useful manual for military surgeons for some years. His sketches and paintings continue to be interesting as works of art and have provided much insight to the nature and treatment of war wounds during the Napoleonic Wars. Before the advent of practical photography, they served as teaching aids in the training of doctors, both civilian and military. Knox, Bell's former surgical assistant in Brussels, succeeded Dr. John Barclay in his extramural school of anatomy in Edinburgh in 1825[N] and was able to persuade the Edinburgh College of Surgeons to establish a museum of comparative anatomy and pathology, becoming its conservator (Fig. 5.4). His face disfigured by smallpox during childhood, Knox was flamboyant, abrasive, and outspoken, and he eventually lost his professional standing.[O] It was Knox, however, who encouraged the purchase of Bell's collection of more than three thousand specimens for three thousand pounds and who arranged its transfer from London. The transfer occurred in two stages, one in October 1825 and the other ten months later, with the collection being first shipped to Leith and then conveyed to Surgeons' Square in four-wheeled spring wagons lent by the Artillery. It consisted of pathological and anatomical specimens, wax casts, firearm missiles retrieved from the living and the dead, and the series of fifteen oil paintings of war wounds made by him after the retreat to Corunna that are now the pride of the museum of the Royal College of Surgeons of Edinburgh (Appendix 3).

In celebration of the quincentenary of the college, in 2005 a slim volume of the paintings and sketches of Sir Charles Bell (of the wounded at Corunna and Waterloo) was published under the title *A Surgical Artist at War* accompanied by Bell's original annotations and followed by a brief commentary on each case, its treatment, and its context from a more modern perspective.[25] There are pictures of people with old or recent wounds of the head, chest, abdomen, scrotum, and limbs, and a most arresting sketch in oil of opisthotonus (a spasm in which the head and body arch backward in extreme hyperextension so that the body is raised off the bed in a bow shape) created from observations of three patients who developed tetanus after head wounds (Fig. 5.5). A set of the Waterloo

---

[N] John Barclay (1758–1826), a former divinity student and preacher who then graduated in medicine at Edinburgh, was a much-respected teacher of anatomy and related topics at 10 Surgeons' Square, his extramural school in Edinburgh, from 1797 until 1825. He gave his collection of comparative anatomy specimens to the Royal College of Surgeons in that city, where they formed the Barcleian Museum. He was interested in veterinary medicine and had a major role in the founding of a veterinary school by one of his pupils at Edinburgh University.

[O] Knox fell from grace with the public and his colleagues because of his involvement with Burke and Hare, two murderers who sold him the bodies of their victims for use in his dissection rooms. Arrested in 1828, Hare turned King's evidence, and Burke was hanged. Knox was not prosecuted, but he resigned from the museum in 1831 and his once popular anatomy classes were soon overshadowed by those offered by Edinburgh University. Unable to obtain a university chair and expelled from various professional societies or organizations, he was ruined professionally. He died in 1862.

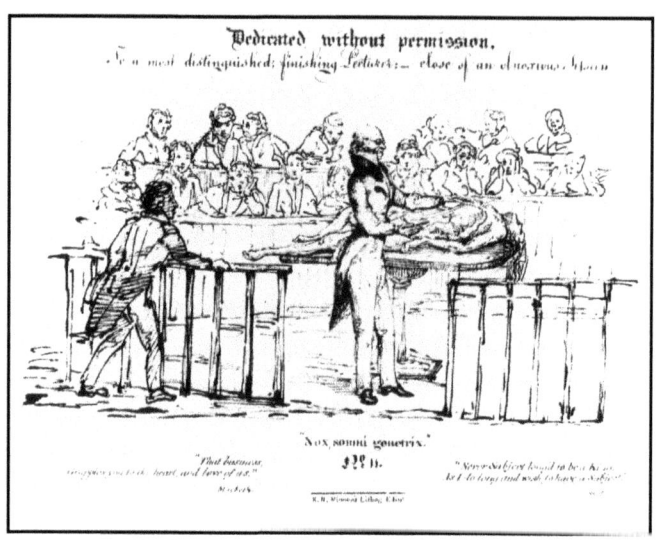

EXECUTION OF WILLIAM BURKE
From an Etching by Walter Geikie. Edinburgh, 1829

Figure 5.4 (Top) Dr. Robert Knox, a former surgical assistant to Bell, is shown dissecting a cadaver during an anatomy lecture in this 1829 lithograph. Knox became involved with Burke and Hare, two murderers who sold him the bodies of their victims for use in his dissection rooms. He was not prosecuted but was ruined by the affair. (Courtesy of the US National Library of Medicine.) (Bottom) The execution of William Burke, who supplied the bodies of his murdered victims to Knox and others for dissection. Etching by Walter Geikie, from Ball JM: *The Sack-'em-up-Men*. Oliver & Boyd: Edinburgh, 1928. (Courtesy of Wellcome Library, London.)

Figure 5.5 Bell retained his clinical interests throughout his professional life. (Top) Opisthotonus in a patient with tetanus. From Bell C: *The Anatomy and Philosophy of Expression as Connected with the Fine Arts.* Murray: London, 1844. (Courtesy of the Wellcome Library, London.) Bell used this illustration in a number of his books. It is based on one of his oil paintings of the wounded at Corunna (item 11; see Appendix 3). (Bottom) Sketch by Bell of a patient with a tumor that occupied the greater part of the mouth, antrum, and nasal cavity and extended so far back that it affected both optic nerves, causing blindness. From Bell C: *Surgical Observations; Being a Quarterly Report of Cases in Surgery; Treated in the Middlesex Hospital, in the Cancer Establishment, and in Private Practice. Embracing an Account of the Anatomical and Pathological Researches in the School of Windmill Street.* Vol. I, Part III, plate X. Longman, Hurst, Rees, Orme, and Brown: London, 1816–1818.

watercolor paintings was given to University College London, where for some years they were used for teaching, but they have since disappeared. Other paintings and sketches—including seventeen large watercolor drawings of the wounded at Waterloo—together with his now-missing sketchbook (which contained forty-five sketches from which the larger watercolor drawings were derived) and a copy of his dissertation on gunshot wounds (published before Waterloo) interleaved with blank pages on which he had added observations relating to the 1815 battle and some of his operative notes (also missing), were presented by his widow to the Army Medical Department and put on display at the Royal Hospital Netley, a military hospital built in the mid-nineteenth century at the suggestion of Queen Victoria and that continued to function as a hospital until it closed in 1958. These paintings were for many years on loan to the Wellcome Trust for the History of Medicine in London (Appendix 3), but they are now back at the Army Medical Services Museum in Aldershot, Surrey (England). They present a dramatic contrast with the usual images and portraits of the glories of war encountered in the museums and art galleries of the world.

REFERENCES

1. Howard MR: Medical aspects of Sir John Moore's Corunna campaign, 1808–1809. *J R Soc Med* 1991; 84: 299–302.
2. Crumplin MKH: The Myles Gibson military lecture: Surgery in the Napoleonic Wars. *J R Coll Surg Edinb* 2002; 47: 566–578.
3. McGrigor J: Observations on the fever which appeared in the army from Spain on their return to this country in January 1809. *Edinb Med Surg J* 1810; 6: 19–32.
4. Hooper R: Account of the diseases of the sick landed at Plymouth from Corunna. *Edinb Med Surg J* 1809; 5: 398–420.
5. Bell C: Letter to George Bell dated 22 January 1809. pp. 138–139. In *Letters of Sir Charles Bell, K.H., F.R.S. L. & E: Selected from His Correspondence with His Brother George Joseph Bell*. Murray: London, 1870.
6. Birbeck E: The Royal Hospital Haslar. From Lind to the 21st century. *J R Nav Med Serv* 2012; 98: 36–38.
7. Bell C: Letter to George Bell dated 3 February 1809 from Haslar Hospital. pp. 139–140. In *Letters of Sir Charles Bell, K.H., F.R.S. L. & E: Selected from His Correspondence with His Brother George Joseph Bell*. Murray: London, 1870.
8. Bell C: Undated letter to George Bell from Leicester Street (London). pp. 140–142. In *Letters of Sir Charles Bell, K.H., F.R.S. L. & E: Selected from His Correspondence with His Brother George Joseph Bell*. Murray: London, 1870.
9. Bell C: Letter to George Bell dated 18 May 1809. pp. 145–146. In *Letters of Sir Charles Bell, K.H., F.R.S. L. & E: Selected from His Correspondence with His Brother George Joseph Bell*. Murray: London, 1870.
10. Bell C: Letter to George Bell dated 27 May 1809. pp. 148–150. In *Letters of Sir Charles Bell, K.H., F.R.S. L. & E: Selected from His Correspondence with His Brother George Joseph Bell*. Murray: London, 1870.
11. Bell C: Letter to George Bell dated 23 May 1809. pp. 146–148. In *Letters of Sir Charles Bell, K.H., F.R.S. L. & E: Selected from His Correspondence with His Brother George Joseph Bell*. Murray: London, 1870.
12. Bell C: *A Dissertation on Gun-Shot Wounds*. Longman, Hurst, Rees, Orme, & Brown: London, 1814.

13. Anon: Art. III. An essay on gun shot wounds. By Charles Bell, Surgeon to the Middlesex Hospital, Lecturer in Anatomy in the Theatre, Windmill-Street, etc. 8vo. Longman & Co, 1815. *British Critics* 1815; 4: 486–493.
14. Blanco RL: *Wellington's Surgeon General: Sir James McGrigor.* Duke University Press: Durham, NC, 1974, pp. 144–161.
15. Kempthorne GA: The Waterloo campaign. *J R Army Med Corps* 1933; 60: 52–58.
16. Matheson JM: Comments on the medical aspects of the battle of Waterloo. *Med Hist* 1966; 10: 204–207.
17. Howell HAL: The British medical arrangements during the Waterloo campaign. *Proc R Soc Med* (section on medical history) 1924; 17: 39–50.
18. Bell C: Letter to Francis Horner, Esq., M.P. from 34 Soho Square, London, dated July 1815. pp. 246–248. In *Letters of Sir Charles Bell, K.H., F.R.S. L. & E: Selected from His Correspondence with His Brother George Joseph Bell.* Murray: London, 1870.
19. Bell C: Letter to George Bell, from Brussels, dated July 1, 1815. pp. 240–243. In *Letters of Sir Charles Bell, K.H., F.R.S. L. & E: Selected from His Correspondence with His Brother George Joseph Bell.* Murray: London, 1870.
20. Kaufman MH: Genealogy of John and Charles Bell: Their relationship with the children of Charles Shaw of Ayr. *J Med Biogr* 2005; 13: 218–224.
21. Grzybowski A, Kaufman MH: Sir Charles Bell (1774–1842): Contributions to neuro-ophthalmology. *Acta Ophthalmol Scand* 2007; 85: 897–901.
22. Pettigrew TJ: Sir Charles Bell, K.H., F.R.S. L. & E. Professor of surgery in the University of Edinburgh, &c. &c. &c. pp. 1–22. In *Biographical Memoirs of the Most Celebrated Physicians, Surgeons, etc. etc. Who Have Contributed to the Advancement of Medical Science.* Whittaker: London, 1839.
23. Watts JC: George James Guthrie, Peninsular surgeon. *Proc R Soc Med* 1961; 54: 764–768.
24. Howard M: *Wellington's Doctors: The British Army Medical Services in the Napoleonic Wars.* Spellmount: Staplehurst, UK, 2002.
25. Crumplin MKH, Starling PH: *A Surgical Artist at War. The Paintings and Sketches of Sir Charles Bell 1809–1815.* Royal College of Surgeons of Edinburgh: Edinburgh, 2005.

6

SWINGS AND ROUNDABOUTS

Over the years, Charles Bell developed a close professional and domestic relationship with the children of Charles Shaw, a lawyer and the county clerk in Ayr, near Glasgow in Scotland. Charles Shaw (1757–1827) was the son of the Reverend David Shaw, minister of Coylton for sixty years, and his wife Marion Dalrymple, daughter of James Dalrymple, sheriff-clerk of Ayrshire.[A] In 1785, Charles married Barbara Wright (1766–1847), whose father was an excise collector at Greenock, a port town by the Firth of Clyde, west of Glasgow. Although some reports indicate that the couple had two daughters and four sons,[1] a privately printed annotated book on the family lists five daughters and nine sons, of whom three became surgeons, three lawyers (with one becoming a Writer to the Signet), one an illustrious soldier, and one a wine merchant; one had no specified occupation.[2] Of the three surgeons, two—John (1792–1827) and Alexander (1804–1890)—worked with Bell at the Great Windmill Street School and became surgeons at the Middlesex Hospital in London, and the third—James—became principal inspector general of the medical department of the Madras Army. Among the daughters, the eldest—Barbara—married George Joseph Bell and the next—Marion (1787 or 1788–1876)—married Charles Bell. The Shaw family thus had a similar clerical background to the Bells, both families achieved prominence in the medical and legal professions, and the two families became intertwined by marriage.

Barbara Shaw was born on 12 December 1785, was brought up in Scotland, and married George Bell on 22 October 1806. Bell wrote to his brother five days before the wedding, realizing that the mail between London and Ayrshire took several days:

> I have managed this very ill—I have miscalculated the time, and missed the opportunity of writing to you on the morning of your marriage, as I intended. Believe me, my dear George, I am more pleasingly anxious about you at this moment than I can express. My own marriage I would enter upon with less satisfaction than what I now feel in thinking that your domestic happiness is secured. Nay, there is something selfish in the idea that if broken down in spirit, if incapacitated by ill-health, if subdued by continued opposition,

---

[A] David Shaw served as Moderator of the General Assembly of the Church of Scotland in 1775.

I could come, like an old soldier, and take up my residence by the fireside of my dear and true brother, and my sister Barbara.[3]

The following year, George wrote to Bell asking if he would take on Barbara's second brother John, then about fifteen, as a pupil. Charles responded (on 17 August 1807): "Make your arrangements as you think fit about young Shaw, and remember that I do not acquiesce merely, but will readily and happily agree to them."[4] As indicated in Chapter 3, John came to London as Bell's house pupil, became his favorite student and then his devoted friend, remained with him for years—coming to serve as superintendent of the dissecting room and lecturer at the Great Windmill Street School—and helped to develop Bell's anatomical museum. It was John who did much of the experimental work that led to Bell's publications on the nervous system. He even accompanied Bell to Brussels immediately after Waterloo to care for the wounded, as discussed in Chapter 5. Two years before he died, John was elected surgeon to the Middlesex Hospital, where Bell was on staff.

Almost immediately after John's arrival, Bell began to focus his attention on the function of the nervous system. On 5 December 1807, he wrote to George,

My new Anatomy of the Brain is a thing that occupies my head almost entirely. . . .

I consider the organs of the outward senses as forming a distinct class of nerves from the other. I trace them to corresponding parts of the brain totally distinct from the origins of the others. I take five tubercles within the brain as the internal senses. I trace the nerves of the nose, eye, ear, and tongue to these. Here I see established connections. Then the great mass of the brain receives processes from these central tubercles. Again the greater mass of the cerebrum sends down processes or crura, which give off all the common nerves of voluntary motion, &c. I establish thus a kind of circulation, as it were. In this inquiry I describe many new connections. The whole opens up in a new and simple light; the nerves take a simple arrangement; the parts have appropriate nerves; and the whole accords with the phenomena of the pathology, and is supported by interesting views.[5]

The next two years were busy. In 1808, he canvassed unsuccessfully for the chair of anatomy at the Royal Academy of Arts (see Chapter 4), and in the following year he was involved in caring for the wounded from Corunna, writing up his experiences in his *Dissertation on Gun-Shot Wounds*, published in 1814 (see Chapter 5). In December 1809, Shaw developed scarlet fever and was looked after by Dr. Maton. Scarlet fever is a complication of bacterial infection (usually in the throat) caused by group A *Streptococcus* and typically is a mild illness, but it may be serious or cause long-term medical complications. The infection is contagious, and Bell himself came down with it soon after and was, in turn, looked after by Herbert Mayo, Matthew Baillie, and William Maton and watched over by Shaw, as well as by Astley Cooper, John Abernethy, and his other friends (Fig. 6.1). He afterwards described his delirium:

> As to the delirium it was never such as you suppose; especially the first nights, it was rather agreeable. A painter, with a look of self-gratulation, seemed to place his piece on an easel; another, with an air of superiority, displaced the first and substituted his own style; a third frowned and terrified the last, until, in rapid succession, I saw the finest pieces of history, the most romantic scenery—banditti, ruins, aqueduct.... Every absurdity of my imagination I observed to have a distinct origin in the impression on the sense. When the light was vivid, the candles and fire burning bright, the truth of sensation corrected all aberrations.[6]

The visual hallucinations that he experienced are common in acute confusional states (delirium), as is the fact that hallucinations tend to be reduced or abolished in the light.

In mid-January, Bell got away from the city, going with Shaw to the country cottage (in Hampstead, then outside London) of his friend Lynn, the Westminster Hospital surgeon, to convalesce, while he continued to mull over the functioning of the nervous system, in which he had now become immersed.[7] When he had recovered, Bell went to Edinburgh to visit George.[8] There, he became increasingly attracted to Barbara's sister, Marion, a woman of great beauty and charm, who already was well acquainted with him through his letters to George. Thus began a frequent correspondence in which he told her of his early struggles in London, about his loneliness, and about how much he had looked to George for warmth and companionship. He told her about his professional and intellectual struggles and also about his pride in being received as a gentleman and a person of achievement based on his own merit. In his loneliness he had learned the value of affection and love. He also wrote to her of God, of the book she had recommended "full of the mild doctrines of the better sort of the Church of England," of his parents, and of how—when looking at her—he was reminded of his mother.[9] As for her, she listened well.

MARRIAGE

In the early months of 1811, he and Marion—almost fifteen years his junior—decided to get married, and he made an offer for a house at 34 Soho Square, close to the Great Windmill Street School and also near the Middlesex Hospital (where he was appointed surgeon in 1814). After much rumination about house pupils (who resided with him), he decided to continue to take them on after his marriage despite objections by Marion and her family, who

Figure 6.1 Friends and foes. (Top, left) Sir Astley Cooper, perhaps the most famous and certainly the most highly paid surgeon in Britain, nominated Bell for fellowship of the Royal Society. Mezzotint by S. Cousins after Thomas Lawrence. (Top, right) John Abernethy, a friend of Bell, was the surgeon who founded the medical school of St. Bartholomew's Hospital, London. Mezzotint 1828 by C. Turner, after C. W. Pegler. (Bottom, left) Sir Anthony Carlisle, the surgeon–anatomist, disliked Bell intensely and beat him to the chair of anatomy at the Royal Academy. Engraving by Robinson after Martin Archer Shee. (Bottom, right) Thomas Wakley, surgeon, social reformer, and founding editor of the *Lancet*, never missed an opportunity to bait Bell and other senior figures of the medical establishment. Stipple engraving by W. H. Egleton after J. K. Meadows. (Courtesy of the Wellcome Library, London.)

did not want strangers in the house. He knew he would otherwise have regrets and blame Marion for them.[10]

Soho Square, much frequented by artists, became increasingly popular with the professional classes toward the end of the eighteenth and the early nineteenth centuries. Several other prominent medical men had houses there, including Anthony Carlisle (whom Bell disliked and who had beaten him to the anatomy chair at the Royal Academy, as discussed in Chapter 4; see Fig. 6-1); the square was also popular with lawyers, dentists, and architects. Thomas Barnes, editor of *The Times*, was at No. 25 from 1837 to 1841. Other houses were turned into offices, and No. 32 contained the library and rooms of the Linnaean Society from 1821 to 1857. Several of the houses in the square came to be occupied by small hospitals later in the nineteenth century—the Hospital for Women at Nos. 29 and 30, and the Dental Hospital of London and the National Hospital for Diseases of the Heart and Paralysis, both at No. 32.[11]

Lord Francis Jeffrey (his old friend and editor of the *Edinburgh Review*) wrote to him from Edinburgh on 4 April 1811, with some insight:

> Not many things in this world could give me greater pleasure than the affectionate tone of your letter and the pleasing picture it holds out to me. You are doing exactly what you should do; and if my approbation is at all necessary to your happiness, you may be in ecstasy. I think all men capable of rational happiness ought to marry. I think you in particular likely to derive happiness from marrying; and I think the woman you have chosen peculiarly calculated to make you happy. God bless you. You have behaved hitherto with admirable steadiness and magnanimity and have earned the confidence of all your friends, as well as the means of enjoyment. I cannot lament your nationality very bitterly, both because it holds of all that is happy and amiable, and because I hope it will give us a chance of seeing you often among us. Besides, when you have Scottish tones and smiles perpetually before you, London will become a sort of Scotland to you. You have but two faults in your character, and I think marriage will go a great way to cure them both. One is a little too much ambition, which really is not conducive to happiness; and the other, which arises, I believe, from the former, is a small degree of misanthropy, particularly toward persons of your own profession. Your wife's sweetness of temper will gradually bring you into better humor with the whole world, and your experience of the unparalleled superiority of quiet and domestic enjoyments to all the paltry troubles that are called splendour and distinction, will set to rights any other little errors that may now exist in your opinions. At all events, you will be delivered from the persecution of my admonitions as it would be a piece of unpardonable presumption to lecture a man who has a wife at home to lecture him at home.[12]

Fortunately, Marion and her mother agreed to his continuing to have house pupils, and on 3 June 1811, Bell and Marion were married. They arrived back in London on June 16 after a two-week honeymoon in the Lake District, and visits to Warwick Castle, Blenheim Palace, and Oxford ("the City of Palaces"),[13] and moved in to Soho Square. His wife now provided him with the companionship that he needed, always trying to lighten his load, enjoying his successes, helping him to cope with frustrations and disappointments, and looking after him. He became the focus of her life, and his interests became hers. Marion gave him a happy and secure home, with her even temper, cheery disposition, and refusal to worry about trifles, and she happily took on other chores, such as noting down his thoughts for his books and papers. They both enjoyed the country and being in the fresh air, and they both enjoyed each other. They were to have no children, and perhaps that drew them closer together.[8] They were united also in their simple faith, a belief in a good and kind God. As Marion remembered years later,

> The first reading in the morning was the Scriptures. He would come to look over what I had read, and I see in my prayer-book his marks on passages of gratitude and praise for the marvelous works of the Most High.[14]

BELL'S CHARACTER AND FRIENDS

Bell's personality and behavior were gradually changing from that of the shy young man who had come to London in 1804. He began to take any disagreements, criticism, or opposition to his views in a more openly personal way, and—as he gained in professional stature—any challenges to his professional beliefs and researches became acrimonious. This has led to his sometimes being portrayed as a difficult person. He was certainly volatile and impatient—like John—and a sensitive soul who was quick to take offense, but he was also easily placated. He was a man of principle who did not hesitate—as he got older—to speak his mind perhaps a little too freely, to give his opinion rather too honestly and without diplomatic qualification, following it would seem in the footsteps of his brother. He believed in his own greatness, and he did not hesitate to say so in his letters to George. He was also sentimental and easily moved by a small kindness or a thoughtful gesture of recognition by others. He remained a very private man, hesitant about revealing his own vulnerabilities, fussy in his clothes and appearance.

Although he could be charming, most of those with whom he remained close were friends from his early days or from Scotland and were men of accomplishment or high position. Among them was Francis Jeffrey (1773–1850), mentioned previously, a judge and a literary critic who founded the influential

Edinburgh Review in 1802 and edited it for twenty-six years[B]; Henry Brougham (1778–1868), its co-founder and a frequent contributor; and Henry Thomas Cockburn (1779–1854), another frequent contributor, a judge and one of the leaders of Scotland's Whig party. Henry Brougham became a member of parliament and eventually Lord Chancellor, and helped—as did Bell—to establish the University of London. He is perhaps best remembered for his defense of Queen Caroline in the House of Lords when King George IV divorced her, and it is largely due to him that Cannes—a sleepy village in the south of France where he bought land and spent much time—attracted others among the British aristocracy to build winter homes there and eventually became a popular resort. His name is also recalled by the carriage that was built to his specifications.[C] Bell collaborated with him on an annotated edition of William Paley's *Natural Theology* (see Chapter 10). John Richardson (1780–1864), also an old friend, became an eminent lawyer in London, living in Fludyer Street (where Bell had also lived), mixing with the literary intelligentsia, and serving as crown agent for Scotland. Then there were William Scott (1803–1871), baronet of Ancrum, and Francis Horner (see p. 66), both politicians, and John Cheyne (1777–1836), a physician whose name remains associated with an abnormal pattern of breathing (Cheyne–Stokes respiration).

Another longtime friend was John Crosse, who—as one of his house pupils and a student at the Great Windmill Street School—lived with the Bells shortly after their marriage. He was made to feel one of the family, playing the flute to Marion's accompaniment on the piano, challenging the master of the house at chess, joining in at dinner parties, sharpening his skills as an artist, and meeting many of the most eminent surgeons of the day.[15] Crosse became a respected surgeon, serving with distinction on the staff of the Norfolk and Norwich Hospital in East Anglia for many years.

PROFESSIONAL WORK

Bell's professional work continued undiminished by illness or domestic change. Despite his other writing commitments, in 1810 he had published his *Letters Concerning the Diseases of the Urethra*, which was to go into three editions as well as into American, German, and Swedish editions.[D] He emphasized that urethral strictures (as opposed to obstruction at the bladder neck) were not spasmodic

---

[B] The magazine ceased publication in 1929.
[C] The brougham was a four-wheeled horse-drawn carriage built to his specifications and named after him, but the name is now applied to a style of automobile body.
[D] Bell considered genitourinary surgery to be part of his responsibility as a general surgeon and thus felt qualified to write about his clinical experience and anatomical observations in this field.

(due to muscle contraction) but rather were caused by narrowing of the urethral caliber due to inflammation, exudation, or congestion. Then, in 1812, he described a previously unrecognized muscle (now called Bell's muscle) in the urinary bladder, attached on each side to the mouth of the ureter.[16] Based on his anatomical dissections, he found that this muscle descends toward the bladder neck, with the muscles from each side joining together and running toward the prostate gland in men. Its function, he concluded, is to assist in contraction of the bladder and, at the same time, to draw down the mouths of the ureters and thereby preserve the obliquity of the ureters as they traverse the bladder wall. This obliquity permits a valve-like effect when the bladder contracts—compressing the ureter during urination so as to prevent backflow of urine to the kidneys.[17]

In 1812, Bell took over Hunter's School at Great Windmill Street, paying James Wilson—its owner—two thousand pounds for it, with the proviso that Wilson could continue to live there and teach (see Fig. 3.1). Part of the money may have come from a small dowry that he received when he married, but it was all that he had. The collection of specimens at the school, combined with his own, gave him a magnificent anatomical museum, later sold to the Edinburgh College of Surgeons (see Chapter 3). He gave his first lecture at the school on October 7 to between eighty and one hundred pupils. As he wrote to George, he had now fulfilled his greatest ambition as a teacher.[18]

In 1813, he was encouraged to become a member of the Royal College of Surgeons by Sir Charles Blicke (1745–1815), an eminent surgeon. His oral examination consisted of a single question—Of what disease did he think Napoleon would die?—to which his response is not recorded, and he was duly admitted. (On the following day, he found out that the charade had been arranged in advance.[19]) Then, in April 1814, he wrote to his brother that he had been elected surgeon to the Middlesex Hospital (Fig. 6.2) with a large majority of the votes.[20]

The Middlesex Infirmary (as it was initially called) had opened in 1745[E] with fifteen beds in two adjacent houses in Windmill Street. Despite financial crises and various irregularities, it gradually expanded, with patients being housed in neighboring houses until 1757, when new premises were built in Mortimer Street at what was then the edge of London. These accommodated sixty-four beds, with additional beds in a separate maternity section. Additional wings were added over the following twenty-five years, but financial difficulties and

---

[E] During the eighteenth century, several of London's major general hospitals were founded to supplement the venerable St. Bartholomew's and St. Thomas' Hospitals, which had large endowments. These included Guy's, the Westminster, St. George's, the London, and the Middlesex Hospitals. They relied largely on philanthropy and subscription.

Figure 6.2 Bell was elected surgeon to the Middlesex Hospital in 1814 and worked there for many years. Both John and Alexander Shaw, his brothers-in-law and devoted assistants, also were elected to its staff. View of the entrance to the hospital. From *The Doctor*, 1837. (Courtesy of the Wellcome Library, London.)

irregularities limited their use. In 1791, Samuel Whitbread, a wealthy brewer, endowed a ward for cancer patients. The Napoleonic Wars provoked further financial burdens, but various economies and the introduction of paying patients (admitted for the treatment of venereal disease) reversed the fortunes of the hospital so that by 1815 it had one hundred and seventy-nine beds and was thriving (Fig. 6.3). It was to this hospital that Charles Bell and later the Shaw brothers were appointed.[F]

The appointment meant a great deal to Bell. He was now enabled to serve a large number of patients, had more opportunities for adding to his museum, and

---

[F] Over the years, the hospital expanded, was rebuilt, and had new wings added, but in 2006 it was amalgamated with the nearby and newly rebuilt University College Hospital, with the tower block there subsequently being named the Middlesex Tower. The Middlesex Hospital building was demolished in 2008.

Figure 6.3 A ward at the Middlesex Hospital by Thomas Rowlandson. From *The Microcosm of London*, 1808. (Courtesy of Peter Vinten-Johansen, PhD.)

was assured a greater flow of students to his school and between his school and the hospital. Indeed, as his correspondence records, he soon had a great many patients, including the Russian general Baron Driesen, referred to him by the physician of the Emperor Alexander I of Russia. The general had been wounded at the battle of Borodino by a ball that lodged in his lower thigh, and Bell eventually had to amputate the limb.[G] The case was interesting, and he published a detailed account of it.[21]

With the end of the Napoleonic Wars, London experienced a building boom and the Middlesex was often filled to capacity, with many of the patients being accident cases. There were also many foreign medical visitors, a number of whom Bell entertained to dinner while his wife—in poor health—rested in Scotland. Among them was Philibert Joseph Roux (1780–1854; chief surgeon at La Charité Hospital and later at the Hôtel-Dieu de Paris), who said of Bell that he was one of the few surgeons in London who performed operations in the French manner—"grace without affectation, and a continued attention to do

---

[G] A German surgeon attempted to dissolve the lead ball by means of mercury poured into the wound, for which he was rewarded by a gift from the Emperor Alexander. However, the wound broke down and the baron suffered excruciating pain, perhaps from particles of mercury that were found to have infiltrated the nerves. Bell eventually amputated the limb in January 1817. More details are provided in *Catalogue of the Museum of the Royal College of Surgeons of Edinburgh*, Part I. p. 30. Neill & Co: Edinburgh, 1836.

every thing, in order to arrive quickly at the termination of the cruel act, which constitutes every surgical operation."[22] Another visitor was Johann Gaspar Spurzheim (1776–1832), the phrenologist.[H] Bell and many others of equal prominence opposed phrenology even though it had wide popular appeal and powerful supporters. Eventually discredited, phrenology nevertheless helped to establish the brain as, in Gall's words, "the organ of mind," and it encouraged the idea that certain functions are localized to specific parts of the brain. Driesen and his aide-de-camp also became close to Bell: The general gave him a pair of silver mugs in addition to his fee as a pledge of friendship on parting, while his assistant supposedly burst into tears.

DEATHS IN THE FAMILY

In November 1816, Robert Bell died. He was the oldest in the family, an advocate and legal scholar, articulate, accomplished, and artistic. It was he who had guided the education of the other brothers when their father died. George took over Robert's lecturing duties at Edinburgh University on behalf of Robert's family for the next two years or so. His correspondence with Charles during the following months reflects the sadness and anxieties that followed the death, although at times Charles could not help but express his own continuing need for George's help: "I cannot tell you how much the circumstance of your undertaking Robert's lectures, and your consequent want of time to help me, retards me. I was wont to depend upon you."[23]

Not long after, on 16 April 1820, Charles' brother John, the surgeon–anatomist with whom he had trained and worked in Edinburgh, died in Rome from heart failure. He had retired three years earlier because of ill health and spent the remainder of his life studying art in Italy. His widow returned to Britain, and her first visit was to Charles and Marion in August. It was a sad occasion, worsened for Charles as he went through his brother's papers, many penciled in as notes in front of paintings and statues, and heard about the last days of his life. The notes were "exactly like his surgical papers—sometimes a large portion written at length, then heads of discourse, then two or three leaves torn out of

---

[H] Born in Germany, Spurzheim was for a time assistant and partner to another German, Franz Joseph Gall (1758–1828), who developed the pseudoscience of phrenology (*phrenos*, mind; *logos*, study). They believed that each aspect of mental function related to a specific cerebral region, that the shape of the skull depended on its contained brain, and that overdevelopment of specific parts of the brain was reflected by corresponding features of the skull. Spurzheim improved on Gall's approach, gave it more of a scientific appearance, and popularized it. Complex topographic maps were produced to facilitate the application of this system by individual practitioners and helped to give an air of scientific precision where none existed.

a book and pasted in, from which he has designed to take an extract, an anecdote, or piece of history in illustration of his subject."[24] His notes of those last years were later published by his widow as *Observations on Italy* and were well received.[25] Charles wondered about writing a biographical piece, and he promised that as "fast as I can I shall write out my sketch of his life and criticism of his works, and send it to you."[24] He did in fact send his biographical sketch, but it was never published—on consultation together, the Bell brothers decided that it was better to allow John's works to speak for him than to reignite the controversies into which he had previously been drawn.[26]

FURTHER PROFESSIONAL ACTIVITY, ADVANCEMENT, AND A QUARREL

In the following years, general surgical practice and teaching did not stop Bell from studying the function of the nervous system, his most famous work. He tried out new ideas on John Shaw, his beloved assistant, and it was Shaw whom Bell always tried to convince before moving ahead to advance some new concept. Much of the experimental work on the nervous system was performed by Shaw, who in 1821 went to Paris, where he met the French physiologist François Magendie (1783–1855) and explained Bell's beliefs. In that same year, Shaw published his *Manual of Anatomy*, which was an outline of the demonstrations he gave to the students in the Great Windmill Street School (Fig. 6.4). It contained an outline of Bell's views on the nervous system, and Shaw gave a copy of it to Magendie. Within a year, an acrimonious dispute arose between Bell and Magendie regarding claims for priority in the discovery of certain facts about the nervous system (discussed in Chapter 7). Whether Magendie was influenced by Shaw's insights remains unclear.

In 1825, Shaw was appointed surgeon to the Middlesex Hospital and he thereupon resigned from the Great Windmill Street School.[1] Two years later, on 19 July 1827, at the age of 35 years, John Shaw died at home from a fever (possibly scarlet fever). Charles was powerfully affected by the loss; for a time, the very thought of John made him tearful, and his sleep was disturbed by nightmares. Shaw's body was buried in Hampstead (London). In March of that same year, there was another loss: Barbara—wife of Bell's brother George—also died and was buried in St. John's Chapel, Prince's Street, Edinburgh.

---

[1] John Shaw gained for himself a reputation as an outstanding orthopedic surgeon and authored a well-known book on the spine: *On the Nature and Treatment of the Distortions to Which the Spine, and the Bones of the Chest, Are Subject*. Longman, Hurst, Rees, Orme, Brown, and Green: London, 1823.

Figure 6.4 Certificate that Mr. John Thompson attended lectures on anatomy, physiology, pathology, and surgery at the Theatre of Anatomy (Great Windmill Street) signed in May 1827 by Bell and also by Herbert Mayo and Caesar Hawkins, to whom Bell had just sold the school. (From the archives of the Royal College of Surgeons of England, MS 0219/2.)

Alexander Shaw now took over the duties of his brother John at Bell's school. He had attended Edinburgh High School and afterwards went on to the University of Glasgow and then as a pupil to the Middlesex Hospital. With the aim of obtaining an MD degree, he was admitted a pensioner (fee-paying student) at Downing College, Cambridge, in 1826, but he left to work at the Great Windmill Street School. He did, however, obtain the license of the Society of Apothecaries in 1827 and soon thereafter became a member of the Royal College of Surgeons. He went on to become assistant surgeon at the Middlesex Hospital in 1836 and surgeon in 1842. Among his published works are various accounts of Bell's researches on the nervous system,[27] and in 1869 he republished, with additions, Bell's *Idea of a New Anatomy of the Brain*,[28] first circulated privately in 1811 (discussed in Chapters 7 and 8).[29] The Shaw brothers were thus both remarkable advocates for Bell, even as controversy came to surround his work on the nervous system in the 1820s and thereafter.

Bell himself was feeling increasing pressure from work, especially in the 1820s. He had just published an essay on the circulation (discussed in Chapter 10) that was of particular importance to him. This was followed by his *Illustrations of the Great Operations of Surgery*, published in parts and then in complete form in 1821, with twenty etched plates of his drawings prepared by the famous engraver Thomas Landseer (1795–1880). It is a most beautiful volume. The cost for the complete work with its engraved plates (seventeen hand-colored in full or in part) was five guineas. Bell was also busy with the demands on his time from his surgical practice. A careful and dexterous surgeon with great technical skill but a seeming reluctance to operate, he was a good and kind man who handled his patients with great gentleness and tried to reassure them with sympathy and understanding before any surgery. Nevertheless, he gave his opinions candidly, clearly, and without pretense. The attention that each patient received reflected the complexity of the case rather than wealth or social standing. As a teacher, he was outstanding and nonthreatening, both at the bedside and in the lecture theater, always explaining to his students the pathophysiological basis of symptoms and the principles of diagnosis and treatment, often with the aid of diagrams that he himself drew to emphasize or clarify a point of interest. He constantly updated and practiced his lectures so that they were concise, informative, and unambiguous. He reminded his students often of

> a fault, which I must presume you have in common with other young men educated to surgery ... an intolerable itch to be doing, and a fondness for operation, and an admiration of what is called bold surgery. I say it is my duty as a teacher to restrain this disposition, not to add to it; but to substitute just and humane feelings.[30]

In other words, he urged restraint and the need to avoid doing something just for the sake of doing it.

In 1824, Bell published *An Exposition of the Natural System of the Nerves of the Human Body*, which contained several of his earlier papers from the *Philosophical Transactions of the Royal Society*, but with subtle alterations that seem to credit him for observations made by others. Bell used this *Exposition* to defend vigorously his claims of priority for certain discoveries regarding the nervous system while denouncing those of rivals, as discussed in Chapter 7.[31]

Bell had written to George in July 1822 of "the weariness which comes on a man . . . from protracted years of imprisonment in London; the tedium that comes over him, the sort of despair of change and entire lack of relaxation and pleasure."[32] With his additional burdens, he began thinking of buying a small place in the country, preferably on the bank of a river where he could get away from the ups and downs of his professional life. In London, he felt confined to

> brick walls and dusty streets; if I make an effort, I cannot, with all my diligence, get out of the noise of wheels. If some miles from town I accidentally stand still, I feel what, perhaps, for months I have not perceived—the absence of din."[33]

He went on about fishing,

> How delightful it is to find yourself, in a spring day, by the side of a stream in the midst of a meadow, the fine sloping hills around you, with their drooping trees and broken woods, with your tackle and rod preparing.[33]

And a little later,

> Then if you enjoy a wlder scene—trees, rocks, and torrents—how delightful to stand in the very middle of the stream. A cloud passes over the sun, and suddenly the bright waters take a frowning darkness. And then is the time—you feel the jerk at your elbow, which none but a fisher can speak of.[33]

Despite the attractions of the country, however, Bell enjoyed mixing with the talented and famous in the metropolis and going to the opera and theater, as is apparent from his letters to George. From them, it is also clear that he was constantly short of money, his income from teaching and clinical practice fluctuating greatly from month to month but generally being less than three thousand pounds a year. In such circumstances, he sometimes had to turn to George for help.

In 1824, he was elected professor of anatomy and surgery at the Royal College of Surgeons in London (Fig. 6.5) and during the following four years delivered a series of lectures there (Fig. 6.6).[J] He was also named Hunterian professor of surgery (also 1825–1828), which required an extended series of further lectures for which an honorarium of fifty guineas was paid. Indeed, he delivered fifteen lectures in each of the first three years, and nine in 1828, following which he declared himself indisposed. The honor and professional recognition of

---

[J] These were the honorific Arris and Gale lectures, which are supported by the Arris bequest and Gale's Annuity for Anatomy Lectures, founded in 1646 and 1655, respectively. Six lectures were delivered annually by the person named professor of anatomy and surgery.

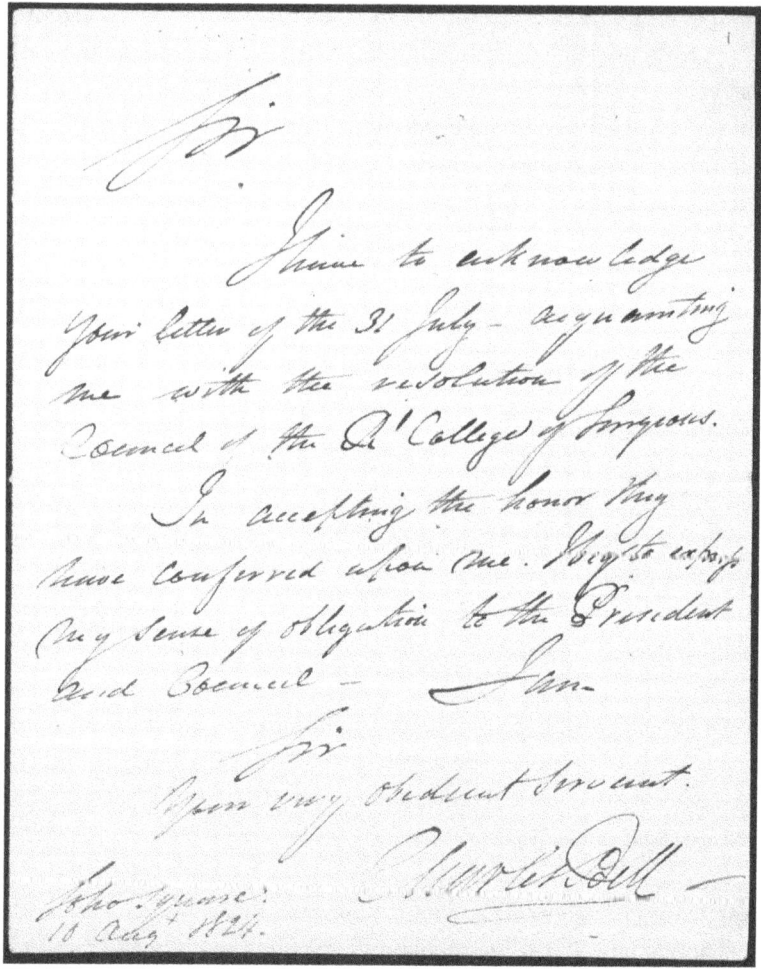

Figure 6.5 Letter from Bell dated 10 August 1824, containing his acceptance of the honor of being appointed Professor of Anatomy and Surgery at the Royal College of Surgeons. (From the archives of the Royal College of Surgeons of England, MS 0219/2.)

these appointments overshadowed the added work and responsibility that they involved. He worked hard to prepare his lectures, hoping to draw the surgeons "back to school again" so that they could keep up to date, and his talks were extraordinarily well received and attended. Artists attended as well as surgeons, seniors as well as juniors, students as well as practitioners. Bell was anxious and apprehensive, especially for his first lecture, but all went well and people came in increasing number to hear him as the course continued so that many people had to be turned away.

It was also in 1824 that Bell quarreled publicly with the popular Astley Cooper, perhaps the leading London surgeon of the period—one of the men

Figure 6.6 Charles Bell lecturing at the Royal College of Surgeons. Lithograph. (Courtesy of the Wellcome Library, London.)

from whom he had sought help when first he came to London—about the treatment of spinal injuries.[34] In December 1823, Cooper had lectured at St. Thomas' Hospital in London on the clinical features of spinal injury, describing cases in which a violent blow to the back had paralyzed the legs. In such cases, he suggested that laminectomy (unroofing of the spinal canal) might be helpful, conceding that in most cases surgery is of no avail and that in many the spinal cord is torn or severed. Uncharacteristically, he referred to those who rejected such an active approach as "foolish" or "a blockhead,"[35] and Bell took these comments personally as his own opinions, relayed in his lectures and writings, were more conservative. Bell was also upset by the "commanding superiority" and the language and tone of Cooper's lecture, which seemed to be "revolting to good taste," and believed that Cooper considered him "deficient [not only] in judgment but in feeling."[36] Not surprisingly, then, he responded angrily in his own lectures at the Great Windmill Street School, subsequently published in book form.[36] He emphasized that in most instances laminectomy served no useful purpose; even in the case of a vertebral fracture with displacement, laminectomy was more likely than otherwise to cause further damage to the spinal cord and increase the risk of infection. He attributed Cooper's comments to "unsatisfactory and inconsequent reasoning" and did not hesitate to criticize him. He also correctly emphasized that in traumatic paraplegia, urinary retention leads to infection (for which he recommended catheterization) and thus to death; other complications included abdominal distension, ventilatory inadequacy, and bed sores.

Fortunately, Bell and Cooper remained on good terms despite their heated exchange. As for the controversy concerning the management of spinal injuries, it continued for years, with some surgeons advocating exploratory laminectomy (but disagreeing on the best time for intervention) and others preferring conservative treatment.[37] Nowadays, modern imaging techniques have made it easier to decide whether to operate.

Bell's intemperate response to Astley Cooper's teachings on the treatment of spinal injuries was contrasted by some in England with his seeming lack of response in the contemporary quarrel with Magendie and Mayo, discussed in Chapter 7.[30] Actually, Bell was very concerned to press his claims against them, but he left it to the Shaw brothers to do so for him.

REFERENCES

1. Kaufman MH: Genealogy of John and Charles Bell: Their relationship with the children of Charles Shaw of Ayr. *J Med Biogr* 2005; 13: 218–224.
2. Shaw J: *The Dalrymples of Langlands*. pp. 109–128. Printed privately: Bath (UK), undated (?1879).
3. Bell C: Letter to George dated 17 October 1806. pp. 82–84. In *Letters of Sir Charles Bell, K.H., F.R.S. L. & E: Selected from His Correspondence with His Brother George Joseph Bell*. Murray: London, 1870.
4. Bell C: Letter to George dated 17 August 1807. pp. 101–103. In *Letters of Sir Charles Bell, K.H., F.R.S. L. & E: Selected from His Correspondence with His Brother George Joseph Bell*. Murray: London, 1870.
5. Bell C: Letter to George dated 5 December 1807. pp. 117–119. In *Letters of Sir Charles Bell, K.H., F.R.S. L. & E: Selected from His Correspondence with His Brother George Joseph Bell*. Murray: London, 1870.
6. Bell C: Letter to George dated 13 January 1810. pp. 166–167. In *Letters of Sir Charles Bell, K.H., F.R.S. L. & E: Selected from His Correspondence with His Brother George Joseph Bell*. Murray: London, 1870.

7. Bell C: Letter to George dated 12 March 1810. pp. 170–173. In *Letters of Sir Charles Bell, K.H., F.R.S. L. & E: Selected from His Correspondence with His Brother George Joseph Bell.* Murray: London, 1870.
8. F.A.: Sir Charles Bell. *Fraser's Magazine* 1875; 11: 88–99.
9. Bell C: Letters to Miss Shaw dated only 1810 and 2 November 1810. pp. 178–181. In *Letters of Sir Charles Bell, K.H., F.R.S. L. & E: Selected from His Correspondence with His Brother George Joseph Bell.* Murray: London, 1870.
10. Bell C: Letter to George dated February 1811. pp. 183–185. In *Letters of Sir Charles Bell, K.H., F.R.S. L. & E: Selected from His Correspondence with His Brother George Joseph Bell.* Murray: London, 1870.
11. Anon: Soho Square: Portland estate. pp. 42–51. In Sheppard FHW (ed): *Survey of London: Volumes 33 and 34: St. Anne Soho.* London County Council: London, 1966.
12. Cockburn, Lord: *Life of Lord Jeffrey, with a Selection from His Correspondence.* Vol. II, pp. 110–111. Lippincott, Grambo: Philadelphia, 1852.
13. Bell C: Letter to George dated 17 June 1811. pp. 188–189. In *Letters of Sir Charles Bell, K.H., F.R.S. L. & E: Selected from His Correspondence with His Brother George Joseph Bell.* Murray: London, 1870.
14. Bell M: Lady Bell's recollections. pp. 405–434. In *Letters of Sir Charles Bell, K.H., F.R.S. L. & E: Selected from His Correspondence with His Brother George Joseph Bell.* Murray: London, 1870.
15. Crosse MV: *A Surgeon in the Early Nineteenth Century: The Life and Times of John Green Crosse M.D., F.R.C.S., F.R.S., 1790–1850.* pp. 37–38. Livingstone: Edinburgh, 1968.
16. Bell C: Accounts of the muscles of the ureters; and their effects in the irritable states of the bladder. *Med Chir Trans* 1812; 3: 171–190.
17. Cockett ATK: Urinary reflux—the physician's dilemma. *Urol Int* 1966; 21: 531–537.
18. Bell C: Letter to George dated 7 October 1812. p. 203. In *Letters of Sir Charles Bell, K.H., F.R.S. L. & E: Selected from His Correspondence with His Brother George Joseph Bell.* Murray: London, 1870.
19. Bell C: Letter to George dated 3 June 1813. pp. 206–207. In *Letters of Sir Charles Bell, K.H., F.R.S. L. & E: Selected from His Correspondence with His Brother George Joseph Bell.* Murray: London, 1870.
20. Bell C: Letter to George dated 7 April 1814. p. 215. In *Letters of Sir Charles Bell, K.H., F.R.S. L. & E: Selected from His Correspondence with His Brother George Joseph Bell.* Murray: London, 1870.
21. Bell C: Case of the Baron Driesen, General in the Russian service. pp. 431–443. In *Surgical Observations; Being a Quarterly Report of Cases in Surgery; Treated in the Middlesex Hospital, in the Cancer Establishment, and in Private Practice. Embracing an Account of the Anatomical and Pathological Researches in the School of Windmill Street.* Longman, Hurst, Rees, Orme, & Brown: London, 1816.
22. Roux PJ: *A Narrative of a Journey to London in 1814; or, A Parallel of the English and French Surgery; Preceded by Some Observations on the London Hospitals.* 2nd edition. p. 26. Cox: London, 1816.
23. Bell C: Letter to George dated 2 March 1818. pp. 259–260. In *Letters of Sir Charles Bell, K.H., F.R.S. L. & E: Selected from His Correspondence with His Brother George Joseph Bell.* Murray: London, 1870.

24. Bell C: Letter to George dated 12 August 1820. pp. 266–267. In *Letters of Sir Charles Bell, K.H., F.R.S. L. & E: Selected from His Correspondence with His Brother George Joseph Bell*. Murray: London, 1870.
25. Avery H: John Bell's last tour. *Med Hist* 1964; 8: 69–77.
26. Bell M: Footnote, p. 267. In *Letters of Sir Charles Bell, K.H., F.R.S. L. & E: Selected from His Correspondence with His Brother George Joseph Bell*. Murray: London, 1870.
27. Shaw A: *Narrative of the Discoveries of Sir Charles Bell in the Nervous System*. Longman, Orme, Brown, Green, and Longmans: London, 1839.
28. Shaw A: Reprint of the "Idea of a New Anatomy of the Brain; submitted for the observations of his friends; by Charles Bell, F.R.S.E." To which are added selections from Letters written by the author to his brother, Professor George Joseph Bell, between the years 1807 and 1821. *J Anat Physiol* 1868; 3: 147–182.
29. Bell C: *Idea of a New Anatomy of the Brain; Submitted for the Observations of His Friends*. Strahan & Preston, 1811. Reprinted in *Medical Classics* 1936; 1: 105–120.
30. Bell C: *Observations on Injuries of the Spine and of the Thigh Bone: In Two Lectures Delivered in the School of Great Windmill Street. The First in Vindication of the Author's Opinions Against the Remarks of Sir Astley Cooper, Bart. The Second on the Late Mr. John Bell's Title to Certain Doctrines Now Advanced by the Same Gentleman*. p. 16. Tegg: London, 1824.
31. Bell C: *An Exposition of the Natural System of the Nerves of the Human Body*. pp. 2–3. Spottiswoode: London, 1824.
32. Bell C: Letter to George dated 1 July [1822]. pp. 275–276. In *Letters of Sir Charles Bell, K.H., F.R.S. L. & E: Selected from His Correspondence with His Brother George Joseph Bell*. Murray: London, 1870.
33. Bell C: Letter to George (?) dated 28 July 1824. pp. 284–287. In *Letters of Sir Charles Bell, K.H., F.R.S. L. & E: Selected from His Correspondence with His Brother George Joseph Bell*. Murray: London, 1870.
34. Silver JR: History of the treatment of spinal injuries. *Postgrad Med J* 2005; 81: 108–114.
35. Cooper A: Surgical lectures. Lecture nineteenth. *Lancet* 1823: 1: 393–398.
36. Bell C: Preface, and Lecture on the spine. pp. iii–xv and 3–31. In: *Observations on Injuries of the Spine and of the Thigh Bone: In Two Lectures Delivered in the School of Great Windmill Street. The First in Vindication of the Author's Opinions Against the Remarks of Sir Astley Cooper, Bart. The Second on the Late Mr. John Bell's Title to Certain Doctrines Now Advanced by the Same Gentleman*. Tegg: London, 1824.
37. Guttman L: Surgical aspects of the treatment of traumatic paraplegia. *J Bone Joint Surg* 1949; 31B: 399–403.

7

IN AND OUT OF THE CENTRAL NERVOUS SYSTEM

Much of what we now take for granted about the nervous system was unknown at the beginning of the nineteenth century. The nervous system is divided into central (the brain and spinal cord) and peripheral portions (the nerves). Nerves conducting impulses to the central nervous system are called afferent nerves and are concerned with sensation; those conducting impulses away from it to the muscles or other effector structures are the efferent nerves. The afferent (sensory) fibers provide input to the brain and spinal cord (in the form of electrical impulses) concerning the external world and about the body itself; efferent (motor) impulses allow the animal to respond to its external or internal environment. The building blocks of the nervous system are nerve cells (or neurons) and supporting cells (glia). Each neuron consists of a cell body, an axon or nerve fiber (that conducts impulses between the cell body and the target with which it connects), and dendrites (short branches coming off the cell body, which receive impulses from other cells).

In the peripheral nervous system, the afferent and efferent nerve fibers are gathered together as individual nerves. The nerves are connected to the spinal cord by anterior and posterior nerve roots. The posterior root has an attached ganglion (the posterior or dorsal root ganglion) that contains the parent cell bodies of the afferent nerve fibers, one branch of which is connected with the periphery and the other with the central nervous system. Parts of the central nervous system are gray from their contained cell bodies of neurons, whereas other parts are white, consisting mostly of nerve fibers. The gray matter lines the surface of the brain and also occurs in various nuclei within its depths; by contrast, it has a butterfly-shaped appearance in the center of the spinal cord.

The human brain consists of the cerebrum (the two cerebral hemispheres); the cerebellum (little brain) at the back of the head, now held to have motor and cognitive functions; and the brainstem, which is structurally continuous with the spinal cord, connecting it and the other parts of the brain with each other, and which maintains or regulates arousal, contains cardiorespiratory and other vital control centers, maintains autonomic functions, and gives rise to the cranial nerves.

Motor and sensory nerves were recognized as early as 300 BC, but their anatomical distinction was unclear and disputed. It was a common belief in the late eighteenth and early nineteenth centuries that input of a sensory stimulus to the central nervous system led to a motor or efferent output generated in the so-called *sensorium commune*, variously located in the white matter of the whole brain, in different parts of the brain, or in the spinal cord and certain parts of the brain.[1] Some scientists thought that it was possible for this sensory–motor conversion also to occur in nerve ganglia or plexuses by means of interconnecting branches.[1] The mechanism involved was unknown.

The early nineteenth century was a time of turmoil. In Europe, the Napoleonic Wars continued, but major defeats of the French in land and sea battles culminated eventually in the battle of Waterloo; in North America, war broke out between the United States and the British and their North American colonies and allies; and in South America the peoples of Argentine, Uruguay, Venezuela, Chile, and Paraguay struggled for independence. In Britain, the monarch's eldest son became Prince Regent in 1811 because of his father's insanity. Despite this unrest, significant developments and discoveries occurred in the natural and biological sciences, including remarkable advances in the understanding of the function of the brain.

Charles Bell—in his introductory comments to a volume on the nervous system that he wrote in 1803—had stated, "There can be no natural division of the nervous system, for it is a whole so connected in function, that no one part is capable of receiving or imparting any sensation, or of performing the operation of the intellect."[2] His views were soon to change. As far back as the sixteenth century, descriptions of the anterior and posterior nerve roots were not accompanied by any appreciation of their physiological significance—of whether and how they differed.[3] In March 1810, Bell described in a letter to his brother George how he had found differences between them in experiments on animals (but did not state which animals he studied or the experimental conditions):

> Experiment 1. I opened the spine and pricked and injured the *posterior* filaments of the nerves—no motion of the muscles followed. I then touched the *anterior* division—immediately the parts were convulsed.
> Experiment 2. I now destroyed the posterior part of *the spinal marrow* by the point of a needle—no convulsive movement followed. I injured the anterior part, and the animal was convulsed.
>
> It is almost superfluous to say that the part of the spinal marrow having sensibility comes from the cerebrum; the posterior and insensible part of the spinal marrow belongs to the cerebellum.[4]

Thus, Bell demonstrated differences between the nerve roots. He described his findings again in 1811 in his *Idea of a New Anatomy of the Brain*, which was printed privately for his friends, with only one hundred copies made.[5] The book—it was more a pamphlet—was short and without illustrations or references, but it has been said to mark the birth of modern neurology and labeled a neurological *Magna Carta*. For unclear reasons, he did not attempt to publish his findings in a scientific journal or make his book more generally available, perhaps because he was concerned that he might be ridiculed or found to be wrong. He believed, however, in disseminating his ideas by teaching and did so freely in his anatomy school. As he had written about his new anatomy to George on 5 December 1807, "My object is not to publish this, but to lecture it—to lecture to my friends—to lecture to Sir Jos. Banks' coterie, to make the town ring with it."[6] In this context, his ideas could mature with time, thought, and discussion.

Bell's small book was more than simply an account of his studies on the nerve roots, but unfortunately this one aspect—because of its fundamental importance and the bitter claims about priority that followed—has diverted attention from the other concepts contained therein. In reality, the book was a critique of the contemporary concept of neurological organization, and its focus was the brain. The scope of the book was so comprehensive that he had mulled over it for several years before having it printed privately. Bell recognized that "the divisions and subdivisions of the brain, the circuitous course of nerves, their intricate connections, their separation and re-union, are puzzling in the last degree, and are indeed considered as things inscrutable." His aim was to offer reasons for believing

> that the cerebrum and cerebellum are different in function as in form; that the parts of the cerebrum have different functions; and that the nerves which we trace in the body are not single nerves possessing various powers, but bundles of different nerves, whose filaments are united for the convenience of distribution, but which are distinct in office, as they are in origin from the brain.[5]

He also discussed the basis for sensation, stressing that "perception is according to the part of the brain to which the nerve is attached."[5] These various points and Bell's subsequent writings on the nervous system are discussed in this and Chapter 8, and his *Idea* is reprinted in its entirety in Appendix 4.

BELL AND THE NERVE ROOTS

It is important to examine what Bell wrote in his original pamphlet about the spinal nerve roots, which repeats and expands on his earlier correspondence with George, because a controversy subsequently arose about priority for the discovery of the differing function of these roots. Only a few sentences bear on this topic:

> On laying bare the roots of the spinal nerves, I found that I could cut across the posterior fasciculus of nerves, which took its origin from the posterior portion of the spinal marrow without convulsing the muscles of the back; but that on touching the anterior fasciculus with the point of the knife, the muscles of the back were immediately convulsed.
>
> Such were my reasons for concluding that . . . every nerve possessing a double function obtained that by having a double root.
>
> The spinal nerves being double, and having their roots in the spinal marrow, of which a portion comes from the cerebrum and a portion from the cerebellum, they convey the attributes of both grand divisions of the brain to every part;[5]

In other words, the nerve roots differed in function. The anterior roots but not the posterior roots connected to the cerebrum via the spinal cord and had motor functions, although Bell thought at this time that they also subserved sensation from the limbs:

> The cerebrum I consider as the grand organ by which the mind is united to the body. Into it all the nerves from the external organs of the senses enter; and from it all the nerves which are agents of the will pass out.[5]

He did not establish the role of the posterior nerve roots, but he thought they connected with the cerebellum[A] and that "they unite the body together, and controul the actions of the bodily frame; and especially govern the operation of the viscera necessary to the continuance of life."[5]

Bell's book did not provide any details of his experimental findings. It has been taken by some to imply that the anterior root is motor, and the posterior sensory. In fact, it suggested that the anterior root contains both motor and sensory fibers, while the posterior root contains fibers from the cerebellum, which Bell thought were autonomic in nature. It was not until several years later that

---

[A] The functions of the cerebellum are still being elucidated, but it seems mainly concerned with regulating balance and coordination of movement as well as having some cognitive functions. Thomas Willis, in the seventeenth century, suggested that it regulated involuntary actions, such as cardiac, respiratory, and gastrointestinal functions, which are today believed to be regulated by the autonomic nervous system. Charles Bell accepted Willis' view. Others in the eighteenth and early nineteenth centuries ascribed sensory (*sensorium commune*) or cognitive functions to it or held that it was involved in sexual drive.

the true differences between the anterior and posterior roots were discovered. This discovery was of fundamental importance and came to be ranked with William Harvey's discovery of the circulation of the blood. Bell was to claim the credit for it, but he was not the only one to do so.

MAGENDIE AND THE NERVE ROOTS

François Magendie (Fig. 7.1), the renowned Parisian experimental physiologist,[B] came to study the functions of the anterior and posterior spinal roots some years after Bell's book had appeared. To this end, he used a litter of eight six-week-old puppies that he had received as a Christmas gift. His experiments involved cutting through the vertebral column (which was not yet fully ossified in these young animals) and then the nerve roots individually, without damaging the adjacent spinal cord. When he cut the posterior roots on one side in the lumbosacral region and then allowed the animal to recover, he found that power was preserved but sensation was lost in the limbs that these roots innervated. Cutting through the anterior roots instead caused flaccid paralysis of the limb without affecting sensation.[C] Cutting both roots caused loss of power and sensation. After replicating these findings, he published his results for all to see in August 1822, concluding correctly that the posterior and anterior roots do indeed have different functions, the former conveying sensory fibers and the latter motor fibers. Thus, Magendie claimed to have discovered the different functions of the anterior and posterior nerve roots.[7]

Magendie is said to have been unaware of Bell's earlier work. The *Idea* was brought out without a date, either on the title page or elsewhere. Alexander Shaw subsequently contacted the printers and was told that one hundred copies were printed at the end of August 1811, and these were dispatched by Bell only to a select group of his friends and colleagues that did not include Magendie.[8] His

---

[B] French physiologist–vivisectionist and physician Magendie became professor of medicine at the Collège de France in Paris. He wrote the first modern textbook of physiology, described an aperture or foramen in the brain, now named after him, that permits the cerebrospinal fluid bathing the brain to circulate freely, and gave an early description of the fluid itself. In addition to his work on the spinal nerve roots, he studied the mechanics of vomiting, the effect of protein deficiency in the diet, the digestive properties of pancreatic juice, the role of the liver in detoxification processes, and the effects of various drugs, including strychnine. His work in demonstrating the presence of sugar in the blood led his assistant, Claude Bernard, to discover that the liver normally produces sugar or a substance (glycogen) readily converted to it.

[C] Magendie's experimental methods were criticized strongly in Britain during the early nineteenth century for their cruelty, involving vivisection without anesthesia (which was then not available), often on a seemingly unnecessary scale. His public lecture–demonstrations in London provoked an outcry and strong anti-vivisectionist sentiment.

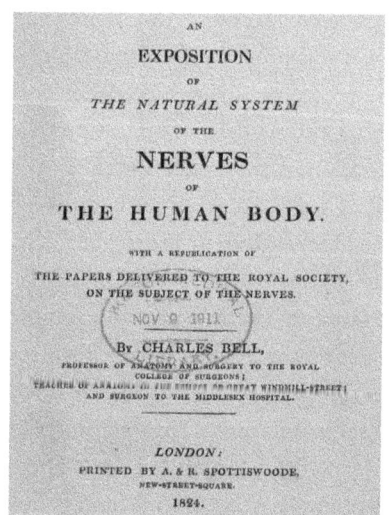

Figure 7.1 Charles Bell (top, left) became embroiled in a priority dispute with François Magendie (top, right) about the discovery of the different functions of the anterior and posterior spinal roots and with Herbert Mayo (bottom, left) about the discovery of the different functions of the fifth and seventh cranial nerves. He republished his earlier papers to encompass their findings (bottom, right). (Images of Bell and Magendie courtesy of the National Library of Medicine, and Mayo courtesy of the Wellcome Library, London.)

essay was first reproduced in its entirety in an American journal in 1812,[9] without Bell's knowledge.[D] The journal was new, obscure, and ceased publication after only one year, so Magendie may well have been unaware of it. However, John Shaw met with Magendie in Paris in 1821, explained Bell's beliefs,[10] and gave him

---

[D] Bell had sent a copy of his *Idea* to William Gibson, one of his former students, who was professor of surgery at the University of Baltimore and went on to gain the chair of surgery at the

a copy of his own just-published *Manual of Anatomy*, which contained an outline of Bell's views on the nervous system. And in April 1822, John Shaw referred publicly to the *Idea* in a paper read before the Medical and Chirurgical Society of London titled "On Partial Paralysis."[11,E] Thus, it may be that Magendie did know of Bell's work and perhaps it piqued his own interest and led him to study the nerve roots. Indeed, early in 1822, Magendie invited Bell to submit a short summary of his work for a prize-medal, although Bell by his own account preferred to focus on new work rather than rehash his publications.[12]

Even if Magendie had known of Bell's earlier work, however, he could not have known in advance that one root was motor and the other sensory, as this was unknown to Bell or anyone else. Nevertheless, as Shaw summed it up, "M. Majendie [sic] ... corroborate[d] some experiments which had been previously made in this country; but of the performance of which M. Magendie does not appear to have been aware."[11] And he later added his belief that "in the investigation of the functions of the several nerves, M. Magendie has not the merit of originality; and that the experiments ... are only following up Mr. Bell's inquiries."[10]

In any event, in a second article that appeared two or three months after the first, Magendie reported on his further findings in animals with strychnine-induced convulsions, as well as on the effects of mechanical and electrical stimulation of the nerve roots.[13] Transection of the posterior roots did not affect the strychnine-induced convulsive movements; when the anterior roots were cut, however, convulsive activity was abolished in the affected limb. In addition to sensory responses (pain), Magendie also found some motor responses—probably reflex in origin—with mechanical stimulation of the posterior roots. Furthermore, electrical stimulation of the anterior roots caused some pain in addition to motor responses.[14] From these findings, he concluded that the posterior roots were *predominantly* sensory, the anterior roots *predominantly* motor.

University of Pennsylvania. Gibson sent the copy on to Dr. Nathaniel Potter, professor of the theory and practice of physic in the College of Medicine in Maryland and editor of the new quarterly *Baltimore Medical and Philosophical Lycaeum*. It was published in the fourth and final issue of that journal with the following letter:

> Dear Sir, I received inclosed in a letter, a few days since from my friend Mr. Chrales [sic] Bell, of London, an essay on the anatomy and physiology of some particular parts of the brain. As this is in all probability, the only copy which has yet reached the United States, it may be gratifying to the faculty to become acquainted as early as possible with the author's views. You are at liberty therefore (if you think proper) to give it a place in your journal.
>
> Yours &c William Gibson, Baltimore, March 13th, 1812

[E] In this paper, Shaw gave the year of publication of the booklet as 1809. The mistake probably arose from his recollection of having, in 1808, transcribed the essay, as discussed in Reference 8.

These conclusions led some (including Bell and Shaw) to misinterpret his views, believing that he now considered each root to have both motor and sensory functions.

THE DISPUTE BETWEEN BELL AND MAGENDIE

Soon after Magendie had published his initial results in 1822, he received a letter from Shaw pointing out that Bell had performed similar experiments some years earlier and had described his findings in his privately circulated book. Magendie responded that he had no knowledge of the book or its content, requested a copy, and read it with the utmost care in the interval between his initial and a second communication. As a result, he was able to include in his second article—in October 1822—a verbatim extract from Bell's work (in English) and concluded with the following comment:

> M. Bell, led by his ingenious ideas regarding the nervous system, was very close to discovering the functions of the spinal roots. . . . However, the fact that the anterior roots are motor while the posterior are specifically sensory seems to have escaped him: Because I established this fact in a positive manner, I make my claims.[13] [this author's translation]

Bell protested, declaring that Magendie had changed his views in his second 1822 paper and now believed that each root had both motor and sensory functions. He was furious when, in April 1823, he read a letter in a daily newspaper, the *Morning Herald*, recounting a conversation overheard in a stage coach in the south of France about certain astonishing experiments on the nerves performed in Paris by Magendie, namely those he regarded as his own.[15] His protestations became more forceful and personal.[F] Bell was not primarily an experimental physiologist but an anatomist and theorist—the very idea of experimenting on living, conscious animals was distasteful to him—and he did not hesitate to comment adversely on Magendie's experimental methods, to suggest that he was incompetent, and to point out that he was French. As he put it,

> In France, where an attempt has been made to deprive me of the originality of these discoveries, experiments without number and without mercy have been made on living animals; not under the direction of anatomical knowledge, or the guidance of just induction, but conducted with cruelty and indifference, in hope to catch at some of the accidental facts of a system, which, it is evident, the experimenters did not fully comprehend.[16]

[F] A detailed account of the controversy, with an annotated bibliography and reprinted texts of the main papers related to it, is found in Cranefield PF: *The Way in and the Way out: François Magendie, Charles Bell and the Roots of the Spinal Nerves*. Futura: Mt. Kisco, NY, 1974.

It is unfortunate that public sentiment against vivisection and animal experimentation was fanned by Bell's remarks. His personal and nationalistic comments harmed only his own cause, for Magendie comported himself with dignity and reserve. The unseemly wrangling did, at least, focus the attention of the scientific community on the functions of the nerve roots and on their central connections.

Bell had not mentioned the spinal roots in any of his own works published between 1811 and Magendie's studies of 1822, but he now quietly revised certain aspects of them, particularly as they related to the motor and sensory functions of the nerves to the face (published in 1821 in the *Philosophical Transactions of the Royal Society*)[17] to incorporate the new results—implying thereby that they were his own—before republishing them in a book in 1824[G] without providing any indication that they had been altered. He attempted in this way to support his claim to having discovered the separate functions of the spinal roots.[18,19] In the introduction to the book, he makes the statement that the anterior roots are motor, the posterior sensory, as if this had been known by him for some time, without any attribution to Magendie, and in a manner that again suggested that the findings were his own.[20] The book itself contains the amended papers, which do not deal specifically with the nerve roots but, by extension of Magendie's work, with the nerves of the face, of the eyeball and orbit. The changes, small but meaningful, are discussed further in the following section.

This behavior is difficult to understand or justify. Even if Bell initially intended to use the results simply to bolster his own claim regarding the function of the brain, he should not have claimed—or appeared to claim—credit for their discovery. In the short term, Bell's cause was helped because the amended publications were accessible and commonly referenced by others in the field; in the long term, his credibility and reputation were damaged as the falsifications came to light and were subjected to public scrutiny. Over the years and continuing well into the twentieth century, correspondence and articles continued to be published, analyzing and reanalyzing the issues and taking a stand on one side or the other concerning priority. They are reviewed elsewhere and need no recapitulation here.[19]

CONCLUSIONS REGARDING THE BELL–MAGENDIE CONTROVERSY

So what is to be concluded? Charles Bell clearly preceded Magendie with his idea of cutting the individual nerve roots separately and in determining that the

---

[G] The papers were republished in his *An Exposition of the Natural System of the Nerves of the Human Body*. Spottiswoode: London, 1824.

anterior—but not the posterior—root had a motor function (although he attributed to it both motor and sensory functions). Even Magendie did not dispute this. It was Magendie, however, who clearly established and characterized the separate functions of these roots, with the anterior roots being motor and the posterior ones sensory. Thus, credit belongs to both men: to Bell for initiating the concept that the two roots have separate functions and the experimental approach (cutting or stimulating the spinal roots in the vertebral canal), and to Magendie for completing the work, defining the actual functions of the roots, and confirming Bell's deductions by experiments that to Bell were unthinkable in their cruelty. In any event, the important concept—that the anterior and posterior roots have different functions—is embodied to this day in the Law of Bell–Magendie, which recalls the contributions of both men.

Bell's work on the fifth (trigeminal) and seventh (facial) cranial nerves also bears on the issue of sensory and motor innervation, but for clarity it is discussed separately in the following section. At least two other scientists also merit consideration. One, Herbert Mayo, is considered in that section. The other, Alexander Walker (1779–1852), was an Edinburgh anatomist who was against animal experimentation, which he believed to be cruel and unable to answer complex physiological problems conclusively.[21] Walker actually suggested in 1809 that the anterior and posterior spinal roots have, respectively, sensory and motor functions (rather than the converse) based on his belief that the anterior portion of the brain was more concerned with sensory functions and the posterior portion with motor activity.[22] Neither concept was correct, and he seems to have arrived at them purely by speculation or—as alleged by some—by faulty plagiarism of Bell's lectures.[23]

Walker later resorted to experiments in frogs to justify his views—his experiments were somewhat similar to those of Bell and Magendie, but his interpretation was quite different. He persevered in his views, claiming credit for discovering the separate functions of the nerve roots. Over time, he has quietly been forgotten, but he believed to the end that he had been robbed of the credit for this discovery. His sentiments are quite apparent from the magnificent title of one of his own books published in 1834: *The Nervous System, Anatomical and Physiological; In Which the Functions of the Various Parts of the Brain Are for the First Time Assigned; And to Which Is Prefixed Some Account of the Author's Earliest Discoveries, of Which the More Recent Doctrine of Bell, Magendie, etc., Is Shewn to Be at Once a Plagiarism, an Inversion, and a Blunder, Associated with Useless Experiments, Which They Have Neither Understood Nor Explained.*

BELL, HERBERT MAYO, AND THE NERVES TO THE FACE

It is now known that the fifth cranial nerve (trigeminal or cranial nerve V) consists of a large sensory portion with an associated ganglion and a smaller motor part without a ganglion. The nerve supplies sensation to much of the face and scalp through its three main branches (the ophthalmic, maxillary, and mandibular branches) and also innervates the muscles of mastication by the mandibular branch. These two parts are analogous to the posterior, ganglionated, sensory roots and anterior, non-ganglionated motor roots of the spinal nerves.[18] By contrast, the seventh cranial nerve (facial nerve or cranial nerve VII; referred to as the portio dura of the nerve by Bell) supplies all the muscles of facial expression.[H]

The distinct function of the nerves to the face was unknown until the work of Bell and of Herbert Mayo (Fig. 7.1), his former student at the Great Windmill Street School and the Middlesex Hospital. Mayo (with Caesar Hawkins) purchased the school from Bell in 1826. He was appointed to the surgical staff at the Middlesex Hospital in 1827 (on the death of John Shaw), and four years later he became professor of anatomy at the newly established King's College London.

Bell, in scientific papers published in 1821 and 1822, had recognized two types of nerves: simple or symmetrical nerves (motor and sensory), which have two roots, and so-called respiratory nerves—innervating muscles involved in respiration, sneezing, coughing, and speaking, including the lips and nostrils—which have one (motor) root and no associated ganglion. His conclusions were based on experimental studies in dogs, asses, and monkeys and on clinical cases that he had encountered. Bell held that the entire fifth nerve had both motor and sensory functions and, similar to a spinal nerve, had two roots, one of which was associated with a ganglion. He believed originally that the nerve carried sensory fibers from the face and also supplied the muscles of both the face and the jaw. The portio dura of the seventh nerve—that is, the facial nerve—he regarded to be a respiratory nerve to the facial muscles, and thus purely motor, although he also reported originally that it sent sensory branches to the cheek.[17,24] He apparently believed, then, that many facial muscles had a dual motor innervation, and which system was used depended on the reason (voluntary or respiratory/instinctive) for their activation. (In fact, many facial muscles do have a dual innervation—for voluntary or emotive movements— not in the nerves that supply them but, rather, in the so-called upper motor neurons that control these nerves.) Bell contrasted the sensory function of the fifth nerve to the predominantly motor function of the facial nerve (one of his respiratory nerves), but he did not compare the function of each of the two roots of the fifth nerve. Thus, any attempt to use these original papers by Bell or his followers as proof that he

---

[H] There are twelve pairs of nerves that arise directly from the brain rather than the spinal cord. Bell referred to these cranial nerves (CN) by the numbering system of Thomas Willis. Their modern equivalents are as follows. The portio dura of CN VII is the facial nerve; the portio mollis of CN VII is actually CN VIII; Bell's VIII nerve is today's CNs IX, X, and XI; Bell's IX nerve is CN XII; and Bell's X nerve is the first cervical nerve.

had preceded Magendie in his understanding of the differing function of the nerve roots is not justified.

In any event, Mayo did not agree with Bell's conclusions as published in the original 1821 article. When branches of the fifth nerve were sectioned, Mayo noted (in 1822) that muscle tone was preserved; when the seventh nerve was cut, by contrast, "the lips immediately fell away from the teeth, and hung flaccid, and the nostrils lost all movement. The sensibility of the lips appeared unimpaired."[25] When the seventh cranial nerve was cut on both sides, Mayo found that all motor functions ceased, indicating that this nerve was a general motor nerve rather than having an exclusively respiratory function. Mayo performed six detailed experiments and concluded that

> the portio dura is a simple nerve of voluntary motion; and that the frontal, infraorbital, and inferior maxillary [branches of the fifth nerve] are nerves of sensation only . . . and from the preceding anatomical details, that other branches of the third division of the fifth, are voluntary [motor] nerves to the pterygoids, the masseter, the temporal [i.e., the masticatory muscles], and buccinator muscles.[1]

Mayo then went through Bell's paper, showing that his conduct of the experiments had confounded interpretation of his results or led to spurious findings. One of the reasons that Bell thought the fifth nerve to be a motor nerve was that, when he cut it in an ass, the animal became unable to eat. Mayo believed the nerve had a sensory function and showed that the animal simply could not feel the food—when food was placed on its tongue, it could indeed eat. He also demolished Bell's attempt to classify nerves as symmetrical and respiratory: "[I]n truth the nerves, which Mr. Bell terms 'respiratory,' do not differ in any important respect, as a class, from those, with which he contrasted them."[25] Modern neuroscientists have tended to discount, also, the classification of nerves as "respiratory" rather than as simply motor in function.

Bell in 1824 changed his mind about the innervation of the face, altered his earlier papers before republishing them,[17,24] and now concluded that the fifth nerve was sensory apart from a branch just to the muscles of the jaw, and the seventh nerve was purely motor (Figs. 7.2 and 7.3). The alterations were subtle, often consisting of the addition or deletion of just a few words, but nevertheless meaningful. They included the now-published findings of both Magendie and Mayo but in a manner that made it seem they were his own findings. For example, with regard to the fifth nerve, Bell had reported in 1821 that it supplied the "muscles of the face and jaws," whereas in 1824 and thereafter the supposedly

---

[1] The buccinator muscle, in fact, is not supplied by the fifth nerve. Mayo is correct about all the other muscles.

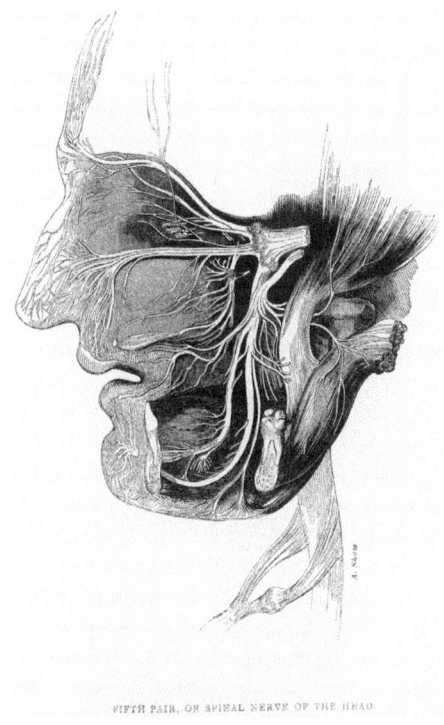

FIFTH PAIR, OR SPINAL NERVE OF THE HEAD.

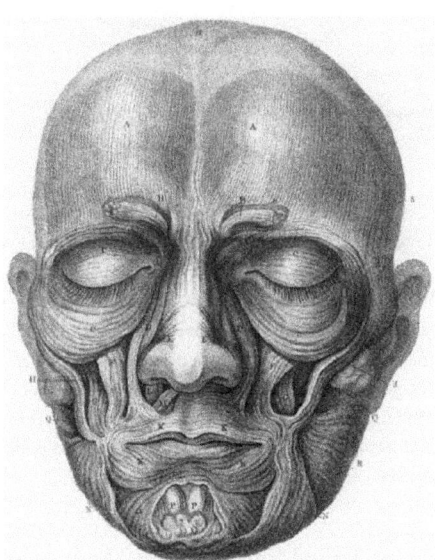

Figure 7.2 (Top) Frontispiece to *Narrative of the Discoveries of Sir Charles Bell in the Nervous System* by Alexander Shaw, 1839, showing the "fifth pair, or spinal nerve to the head." The engraving is by Shaw. (Courtesy of the Wellcome Library, London.) (Bottom) Muscles of the face. (From Bell C: *Essays on the Anatomy of Expression in Painting.* p. 59. Longman, Hurst, Rees, and Orme: London, 1806.)

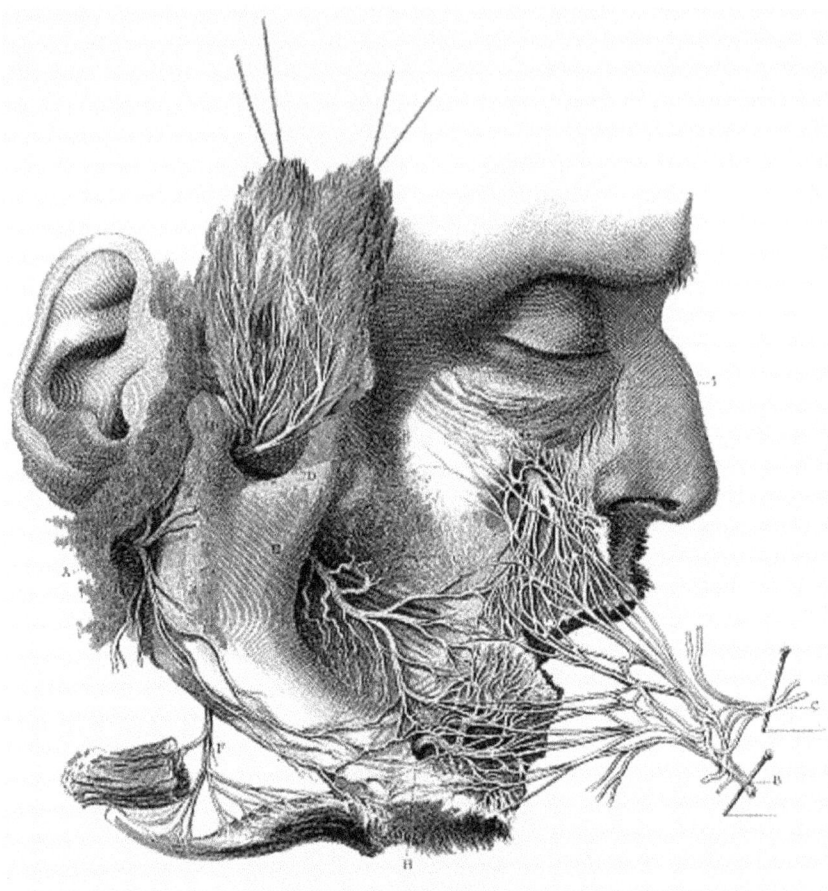

Figure 7.3 The nerves of the face. The superficial nerves of the face are deflected, and the distribution of the motor branches of the third division of the fifth cranial nerve is exposed. *A,* The facial nerve coming out of the stylomastoid foramen; its principal branches are cut and folded forward. *B,* The trunk of the seventh cranial nerve, dissected off the face and pinned out. *C,* Branch of the third division of the fifth cranial nerve, which joins a plexus of the facial nerve before the ear. *D,* The masseter muscle is dissected from the jawbone and lifted up to show branches of the fifth nerve going to this muscle. *D–F,* Motor branches of the fifth nerve. *C* and *G–I,* Sensory branches of the fifth nerve that join branches of the facial nerve in their distribution. Plate VIII. (From Bell C: On the nerves of the face; Being a second paper on that subject. *Philos Trans R Soc Lond* 1829; 119: 317–330.)

same paper stated that it supplied "the muscles of the jaws." Similarly, his earlier comment that the "branches [of the fifth nerve] are profuse to the muscles which move the lips upon the teeth" was changed to "its branches are profuse to the lips."[18,26] It seems, then, that Bell did not correctly describe the functions of the fifth and seventh cranial nerves until two years after Herbert Mayo had given an accurate description of them.

Acrimonious correspondence followed in the columns of the medical and scientific journals concerning Bell's quarrel with both Magendie and Mayo, in which Alexander Shaw advocated on behalf of Bell. At one point, it became so offensive that Bell publicly affirmed that he was not a behind-the-scene correspondent in the dispute. Nevertheless, Mayo was still able to write in his 1827 textbook,

> "But when thus sharing the claim to these discoveries between M. Magendie and myself, I should in justice state that the experiments in each case were but improvements on those which Mr. Bell had previously performed. Mr. Bell's ingenious Essays upon the nerves . . . have the merit which belongs to originality, even when only partially successful in eliciting truth.[27]

Mayo's generosity did not help, and over the years he came to be discredited, robbed of the recognition that he deserved, in part because of published aspersions on his character. By 1833, he was becoming more angry and stated in a footnote to his textbook,

> The reader, who may take a critical interest in this inquiry, should not omit to read Sir C. Bell's *first* essay, printed in the Philosophical Transactions for 1821, in which his opinions are distinctly stated to be such as I have here represented them. In all his subsequent writings on this subject (writings published after the first part of my anatomical and physiological commentaries had in the mean time appeared) Sir C. Bell, retaining the terms which he had employed to explain his theory . . . has substituted for his original opinions my conclusions, I regret to say without acknowledging the source from whence he derived them. It is very painful to me for many reasons to speak in terms of censure of Sir C. Bell; but in justice to myself I am compelled to make the foregoing statement.[28]

In 1836, Mayo became professor of physiology and pathological anatomy at King's College London but, in that same year, he applied unsuccessfully for the just-vacated chair of general anatomy at the rival University College. This action created much ill-feeling and forced his resignation from King's. Illness caused him to withdraw from his duties at the Middlesex Hospital in the early 1840s, and he became physician in a hydropathic establishment in Germany, where he eventually died. His perceived disloyalty to King's College, regarded as evidence of his overarching ambition and self-interest, together with his later involvement in mesmerism and his absence from Britain, helped to ruin his professional standing and credibility.[29] Over this same time, Bell's supporters built up a narrative to undermine Mayo just as it was coming to be accepted that Bell and Mayo deserved equal credit. The deciding factor is said to have been

Alexander Shaw's retelling of the affair, with the texts of Bell's papers—quietly amended—made publicly available.[30] Mayo was painted as a jealous colleague, an ungrateful and disloyal former student who stole his master's ideas, an influential physiologist who misused his position of authority, and an untrustworthy man without personal integrity (perhaps based on his behavior in the Robert Lee affair).[j]

Mayo, for his part, initially did not engage in personal attacks and refused to respond to the allegations against him.[29] He thought, quite simply, that the facts spoke for themselves and he confined himself to criticizing Bell's scientific work until it was too late. When finally he labeled Bell a plagiarist, he went too far in his attacks on a man regarded by many as an elder statesman of the profession.[29] And then, troubled by illness, abandoned by his wife, ridiculed by certain medical journals for his support of mesmerism, he lost all scientific credibility, his claims for priority were rejected, and he was unable to regain the ground he had lost.

Mayo is today a forgotten man, and Bell is usually credited with distinguishing between the functions of the fifth and seventh cranial nerves.

AND SO TO THE SENSE OF SMELL

A brief aside on the sense of smell is warranted. Uncertainty existed in the early nineteenth century concerning the cranial nerve that mediated this sense. Bell believed that the first cranial nerve was responsible,[31] but he thought that some of its fibers joined the fifth nerve—"the 1st nerve must have branches of the 5th united with it"—as indicated in the *Idea*.[s] François Magendie, by contrast, held that it was the fifth nerve, based on a series of experiments in animals.[32] Magendie's experiments were criticized by Shaw[28] and received little clinical confirmation in the medical press. The conclusions of Bell, by contrast, were supported by reports of persons with absent or damaged olfactory nerves who were unable to smell, or with lesions of the fifth nerve who retained the sense of smell. However, irritant chemicals have since been shown to stimulate free nerve endings of the fifth nerve located within the lining of the nose, producing such sensations as burning and stinging. Again, therefore, credit goes to both investigators. A detailed account is provided elsewhere.[33]

---

[j] Robert Lee, an obstetrician, was preparing a paper for submission to the *Philosophical Transactions of the Royal Society* and showed his dissections to several scientists, including Mayo. Mayo initially wrote in strong support of Lee's findings, but a few days later appeared less sure, asking for samples for microscopic examination and subsequently requesting the return of his supporting letter. The affair is discussed by Manuel DE: Robert Lee, the uterine nervous system and a wrangle at the Royal Society 1839–1849. *J R Soc Med* 2001; 94: 645–647.

BELL AND PRIORITIES IN SCIENTIFIC DISCOVERY

There is a certain ugliness or unseemliness in Bell's struggles to claim priority as the originator of a scientific discovery as profound as that of the separate functions of the anterior and posterior nerve roots or of the fifth and seventh cranial nerves. Yet disputes of this sort are frequent in the history of science and have involved such illustrious figures as Galileo Galilei, Isaac Newton, Robert Hooke, and Antoine Lavoisier. Disputes have occurred in the physical and biological sciences, mathematics, astronomy, medicine, and indeed in all the disciplines of learning. Such struggles continue to this day in science and other disciplines. The award of prizes, often of great value, has only promoted such conflict, as has the naming of processes, laws, or diseases after their so-called discoverer. Competition for funding has added to the acrimony. Prestige, recognition, title, fame, and financial support are important rewards, as necessary to scientists as to others.

Discovery rarely occurs as an isolated event at one moment in time. It is a continuous process based on what has gone before. Often, however, the recognition of preceding events is perceived as detracting from the achievement of making a "new" discovery. Some authors have attempted to distinguish between discoveries that were predictable from existing theories and those that were unpredictable.[34] For unpredictable discoveries, one immediate problem is to determine the criteria that have to be met before concluding that something novel has been discovered. Such criteria will affect the determination of when the discovery was made. The process of discovery involves the recognition of an anomaly, of a discrepancy between observed and expected findings based on contemporary beliefs, of a phenomenon requiring explanation. This then leads to further study. Often a discovery is made by more than one person at approximately the same time, providing an even more fertile area for dispute. An additional problem relates to whether credit for a discovery belongs to the originator of a novel belief or idea, the experimentalist who establishes that the idea is correct, or both. Application of the discovery to previous knowledge provides a new dimension to that knowledge. Charles Bell, by his work, did indeed suggest a new way at looking at the nervous system. He was incorrect in many details of his original conception, but his experimental approach was adapted by others—more accomplished experimentalists—and led to advances that were remarkable, that allowed the nervous system to be looked at in new ways and began to make sense of it.

When the same discovery is made simultaneously but independently by more than one person, each of the involved parties sometimes generously credits the other, as exemplified by Charles Darwin and Alfred Wallace and their theory of evolution by natural selection (see Chapter 10).[35] It is noteworthy that both Magendie and Mayo tried to share the credit for their discoveries with Bell, but in each case Bell viewed this as a challenge to his own contribution and would have none of it. Bell was a curious man—insecure, shy, ambitious, impatient, ill-tempered even—who must bear the responsibility for surreptitiously amending and massaging his prior publications to conform with the experimental findings of Magendie and Mayo. The Shaw brothers—who did much to discredit his rivals as they pushed his claims—probably truly believed that Bell deserved recognition for his originality and that others had stolen his intellectual property and the credit that was his due. Thus, Alexander Shaw did not hesitate to promote his cruel and unjust attacks on the character of Herbert Mayo.

Bell's claims to originality were undermined by his failure to publish his findings. If he had done so, he would have received proper credit, the dispute over priority may not have occurred, and subsequent work by Magendie and others would probably have occurred sooner. He believed that he could make his work known to others by lecturing, and he did so freely at his school in Great Windmill Street. Yet he wrote liberally and easily on other topics and had an extensive and impressive bibliography. It seems possible that his *Idea* was just a poorly formulated concept, a draft to be worked on while he refined his ideas and had feedback from friends and colleagues, feedback that he never received. Laid aside, it was not picked up again for more than ten years, until Magendie's experimental work was published.

In the last analysis, it is the historians of science who determine the worth of an individual's contribution.[35] Part of that responsibility is to determine whether unacceptable behavior has sullied the worth of that contribution. In Bell's case, it is difficult to escape the conclusion that he republished his earlier work with subtle alterations that incorporated the findings of Magendie and Mayo without attribution and with the aim of suggesting that the findings were his own. Perhaps he really thought that they were, although this is difficult to believe. In any event, he aggressively asserted his own claims and denounced those of his rivals. Such behavior is disappointing from anyone, but especially from a man with many other, real achievements to his credit.

REFERENCES

1. Clarke E, Jacyna LS: *Nineteenth Century Origins of Neuroscientific Concepts.* pp. 105–106. University of California Press: Berkeley, 1987.
2. Bell C: Introductory view of the nervous system. p. 346. In Bell J, Bell C: *The Anatomy of the Human Body.* Longman, Hurst, Rees, Orme, and Brown: London, 1811.
3. Keele KD: *The Anatomies of Pain.* p. 64. Charles C Thomas: Springfield, IL, 1957.
4. Bell C: Letter to George dated 12 March 1810. pp. 170–173. In *Letters of Sir Charles Bell, K.H., F.R.S. L. & E: Selected from His Correspondence with His Brother George Joseph Bell.* Murray: London, 1870.

5. Bell C: *Idea of a New Anatomy of the Brain; Submitted for the Observations of His Friends.* Strahan & Preston, 1811. Reprinted in *Medical Classics* 1936; 1: 105–120.
6. Bell C: Letter to George dated 5 December, 1807. pp. 117–119. In *Letters of Sir Charles Bell, K.H., F.R.S. L. & E: Selected from His Correspondence with His Brother George Joseph Bell.* Murray: London, 1870.
7. Magendie F: Expériences sur les fonctions des racines des nerfs rachidiens. *J Physiol Exp Pathol* 1822; 2: 276–279.
8. Shaw A: Reprint of the "Idea of a New Anatomy of the Brain; Submitted for the Observations of His Friends; by Charles Bell, F.R.S.E." To which are added selections from letters written by the author to his brother, Professor George Joseph Bell, between the years 1807 and 1821. *J Anat Physiol* 1868; 3: 147–182.
9. Bell C: Idea of a new anatomy of the brain. *Balt Med Phil Lycaeum* 1811; 1: 303–318 [The actual publication date was 1812, but the journal was dated 1811.]
10. Shaw J: Remarks on M. Magendie's late experiments upon the nerves. *Lond Med Physical J*; 1824, 52: 95–104.
11. Shaw J: An account of some experiments on the nerves; by M. Magendie. With some observations. *Lond Med Physical J* 1822; 48: 343–352.
12. Bell C: Letter to George dated 10 June 1822. p. 275. In *Letters of Sir Charles Bell, K.H., F.R.S. L. & E: Selected from His Correspondence with His Brother George Joseph Bell.* Murray: London, 1870.
13. Magendie F: Expériences sur les fonctions des racines des nerfs qui naissent de la moelle épinière. *J Physiol Exp Pathol* 1822; 2: 366–371.
14. Cranefield PF: *The Way in and the Way out: François Magendie, Charles Bell and the Roots of the Spinal Nerves.* pp. 38–39, 52–53. Futura: Mt. Kisco, NY, 1974.
15. Bell C: Letter to George dated 5 April, 1823. pp. 279–280. In *Letters of Sir Charles Bell, K.H., F.R.S. L. & E: Selected from His Correspondence with His Brother George Joseph Bell.* Murray: London, 1870.
16. Bell C: *An Exposition of the Natural System of the Nerves of the Human Body.* pp. 2–3. Spottiswoode: London, 1824.
17. Bell C: On the nerves; Giving an account of some experiments on their structure and functions, which lead to a new arrangement of the system. *Philos Trans R Soc Lond* 1821; 111: 398–424.
18. Waller AD: The part played by Sir Charles Bell in the discovery of the functions of motor and sensory nerves (1822). *Sci Progr* 1911; 6: 78–106.
19. Cranefield PF: *The Way in and the Way out: François Magendie, Charles Bell and the Roots of the Spinal Nerves.* Futura: Mt. Kisco, NY, 1974.
20. Bell C: *An Exposition of the Natural System of the Nerves of the Human Body.* pp. 40–41. Spottiswoode: London, 1824.
21. Rice G: The Bell–Magendie–Walker controversy. *Med Hist* 1987; 31: 190–200.
22. Walker A: New anatomy and physiology of the brain in particular, and of the nervous system in general. *Arch Universal Sci* 1809; 3: 172–179.
23. Boring EG: *A History of Experimental Psychology.* 2nd edition. p. 45. Appleton-Century-Crofts: New York, 1957.
24. Bell C: On the nerves which associate the muscles of the chest, in the actions of breathing, speaking, and expression. Being a continuation of the paper on the structure and functions of the nerves. *Philos Trans R Soc Lond* 1822; 112: 284–312.

25. Mayo H: Experiments to determine the influence of the portio dura of the seventh, and of the facial branches of the fifth pair of nerves. *Anatomical and Physiological Commentaries.* Part 1, pp. 107–120. Underwood: London, 1822.
26. Bell C: *An Exposition of the Natural System of the Nerves of the Human Body.* Spottiswoode: London, 1824.
27. Mayo H: *Outlines of Human Physiology.* p. 240. Burgess and Hill: London, 1827.
28. Mayo H: *Outlines of Human Physiology.* 3rd edition. pp. 192–193. Burgess and Hill: London, 1833.
29. Bradley J: Matters of priority: Herbert Mayo, Charles Bell and discoveries in the nervous system. *Med Hist* 2014; 58: 564–584.
30. Shaw A: *Narrative of the Discoveries of Sir Charles Bell in the Nervous System.* Longman, Orme, Brown, Green, and Longmans: London, 1839.
31. Bell C: The first pair of nerves; or, olfactory nerves. pp. 84–85. In *The Anatomy of the Human Body: Volume III. The Nervous System.* Collins and Perkins: New York, 1809.
32. Magendie F: Le nerf olfactif est-il l'organe de l'odorat? Expériences sur cette question. *J Physiol Exp Pathol* 1824; 4: 169–176.
33. Doty RL: Intranasal trigeminal chemoreception: Anatomy, physiology, psychophysics. pp. 821–833. In Doty RL (ed): *Handbook of Olfaction and Gustation.* Dekker: New York, 1995.
34. Kuhn TS: Historical structure of scientific discovery. *Science* 1962; 136: 760–764.
35. Merton RK: Priorities in scientific discovery: A chapter in the sociology of science. *Am Sociol Rev* 1957; 22: 635–659.

8

THE ORGANIZATION OF THE NERVOUS SYSTEM

Bell came up with a number of original concepts concerning the organization and operation of the nervous system in health and disease in addition to the work discussed in Chapter 7. His work was recognized in 1829, when he received the Gold (Royal) Medal of the Royal Society for his discoveries relating to the nervous system. The medal was created by King George IV and first awarded in 1826; until 1964, only two medals were awarded annually, in the physical and biological sciences, respectively.[A] Bell's published papers and textbooks also provide numerous case histories that give insight to the clinical approach of his day and contain the first descriptions of various neurological diseases, as discussed in Chapter 9.

THE BRAIN AND SPINAL CORD

In the eighteenth and early nineteenth centuries, the brain was held to function as an integrated whole, without specific functions being localized to discrete cerebral regions. This concept evolved into one in which major regions of the brain were endowed with certain specific functional attributes throughout their territory, and these regions were thought to work together to permit the integrated action of the brain as a whole. This led—in the late nineteenth century and most of the twentieth century—to theories of punctate localization of specific functions within the brain and within each of its major regions, especially in the cerebral cortex. It is now believed that there are functional networks of brain cells having specific functions and connecting multiple areas of the brain. Through these networks, the dynamic interaction of different cortical and subcortical areas with multiple interconnections allows specific functions to proceed. Spatially distinct brain regions are thus activated during performance of specific tasks.

Bell believed that the anatomical complexity of the brain had been ignored. In contrast to widely held views that the entire brain made up the *sensorium commune*, he came to emphasize that different parts of the brain have different and

---

[A] The medal in the physical sciences for 1829 went to Eilert Mitscherlich, a chemist, for his work relating to the laws of crystallization and the properties of crystals.

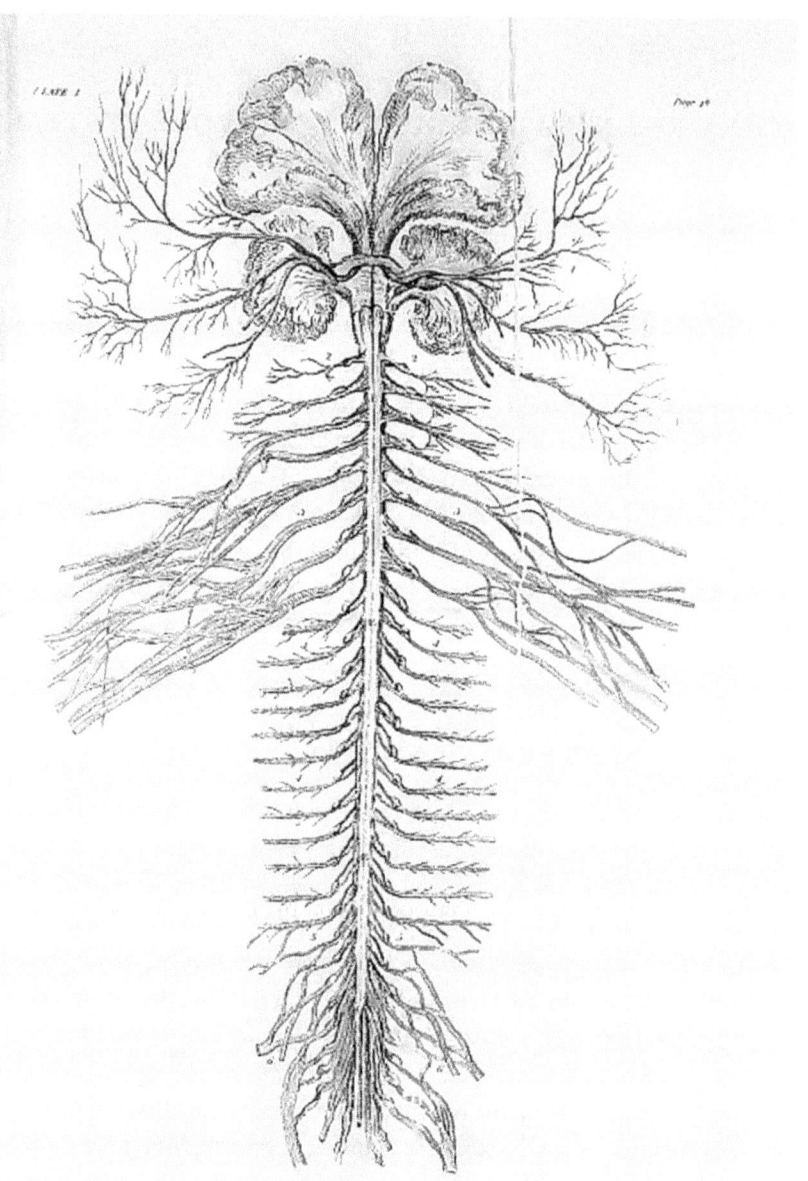

Figure 8.1 The nervous system as depicted by Bell. *A*, cerebrum; *B*, cerebellum; *E*, spinal cord; the different nerves are numbered. (Plate I from Bell C: *An Exposition of the Natural System of the Nerves of the Human Body.* Spottiswoode: London, 1824.)

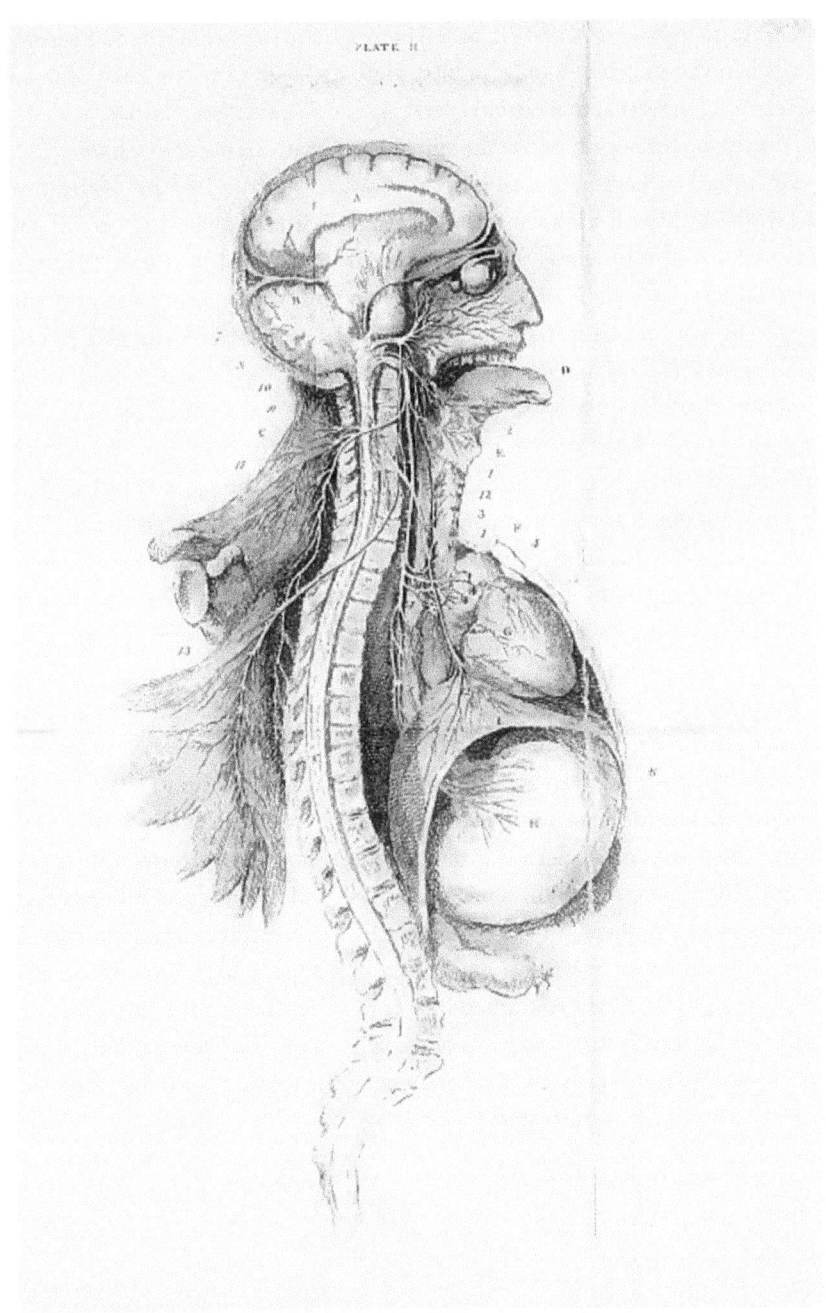

Figure 8.2 The respiratory system of nerves as envisioned by Bell and discussed on pages 107 and 108. *A*, cerebrum; *B*, cerebellum; *C*, spinal cord; *D*, tongue; *E*, larynx; *F*, bronchi; *1*, vagus nerve, passing to the larynx, lungs, heart, and stomach; *2*, superior laryngeal branches of vagus nerve; *3*, recurrent laryngeal branch of vagus nerve; *12*, phrenic nerve to diaphragm; *13*, nerve to serratus anterior (Bell's nerve). (Plate II from Bell C: *An Exposition of the Natural System of the Nerves of the Human Body*. Spottiswoode: London, 1824.)

specific functions, and he distinguished between the cerebrum (the two cerebral hemispheres) and cerebellum (the little brain at the back of the head). He was concerned to show that these have separate functions, and this—in part at least—was the reason for his interest in the spinal roots,[1,2] as discussed in Chapter 7.

He initially accepted that the cerebrum was "the grand organ" of the mind into which "all the nerves from the external organs of the senses enter" and from which "all the nerves which are agents of the will pass out."[3] He thus believed it was specialized for both sensory and motor functions and connected with the anterior half of the spinal cord and the anterior roots of the spinal nerves. It contrasted with the cerebellum (Figs. 8.1 and 8.2), to which he attributed the regulation of autonomic activity ("the secret operations of the bodily frame") and connections with the posterior half of the spinal cord and the posterior roots.[3] In the 1820s, however, once the sensory functions of the posterior roots were established (by Magendie), Bell suggested that the cerebellum was the reception point for sensory information, and that the more anterior parts of the brain were concerned primarily with movement. In an analogous manner, he suggested that the different regions of the spinal cord had different functions.

The spinal cord had traditionally been divided on each side into a posterior column (from the median fissure to the posterior roots), lateral column (between the posterior and anterior roots), and anterior column (between the anterior roots and the anterior median fissure). Based on the different functions of the nerve roots, he thus came to believe that the posterior columns of the spinal cord were the main sensory pathway and the anterior columns were the main motor pathway in the spinal cord.[4] This view remained the prevailing view for much of the next quarter-century. It only began to change after Brown-Séquard, in Paris, examined in animals the effect on sensory functions of selective lesions in different parts of the spinal cord and showed thereby that certain sensory fibers crossed from one side of the spinal cord to the other and then ascended in the anterolateral region.[B]

[B] Charles-Édouard Brown-Séquard (1817–1894) was professor of medicine at the Collège de France in Paris. He correctly identified the sensory pathways in the spinal cord, emphasized the role of functional processes in the integrative actions of the nervous system, discovered the vasomotor nerves (which control the caliber of blood vessels), showed that the adrenal glands are essential to life, and suggested that organs have internal secretions (hormones). He is thus regarded by many as the father of experimental endocrinology, and it was he who initiated hormone replacement therapy. My biography of him (Aminoff MJ: *Brown-Séquard: An Improbable Genius Who Transformed Medicine*) was published by Oxford University Press in 2011.

Bell later changed his views on the cerebellum once more in the light of advancing knowledge and his own clinical experience,[5] concluding that it did not seem to be concerned directly with volitional motor or sensory activity:[C]

> We have no distinct and well-marked cases of disease in the substance of the cerebellum, such as we possess of disease in the cerebrum; and on the whole it does not appear to stand in direct relation to the motions of the frame, or to the common sensibility.[5]

It is clear from Bell's 1811 book—the *Idea*—and subsequent writings that he was attempting to establish certain major neurological systems (a motor–sensory system—later evolving into separate motor and sensory systems—and what would now be called an autonomic system) and, by tracing nerves with different functions into the spinal cord and brain, to determine their central representation and thus the function of different central structures. Once this was established, he could, by tracing the origin of other nerves, determine their function. In other words, he believed that differences in function at the periphery could be extended to different levels of the central nervous system and back. This was indeed a visionary and novel approach for which Bell deserves much credit, but it depended on the ability to trace the connectivity of different parts of the nervous system. Unfortunately, the technology of the period was not up to the task.[1]

Bell was against animal experimentation involving vivisection because of its seeming cruelty and out of concern that the results might be misleading; instead, he believed that any experiments should be guided by "inference from the anatomical structure" and directed "only to the verification of the fundamental principles on which the system is founded."[6] Using his own approach, he hoped to bring some order into contemporary understanding of the operation of the nervous system. For Bell, at least initially, the precise functions of the anterior and posterior spinal nerve roots were therefore less important than the fact that they had different functions, representing different neurological systems. Unlike Magendie, he seemed less concerned about determining the details of these differences.

THE PERIPHERAL NERVES

It followed from the discovery of the differing function of the nerve roots that there are different types of nerve fibers, which normally conduct in only one direction. As Bell noted,[3] individual peripheral nerves actually contain bundles of

---

[C] Curiously, certain medical historians are critical of Bell for his changes of view. To the present author, however, it is entirely reasonable that views should evolve to take account of advances in the field, even if this complicates matters for historians.

nerve fibers with different functions depending on the different parts of the brain from which they arise and the nerve root that they traverse, namely "the nerves of sense, the nerves of motion, and the vital [autonomic] nerves." Regarding the sensory fibers, a prevailing doctrine of the times was that these differed "from each other only by delicacy of structure, and by a corresponding delicacy of sensation, that the nerve of the eye, for example, differs from the nerves of touch only in the degree of its sensibility." Bell concluded, however, that sense organs are specialized to receive only one form of sensory stimulus, and he subsequently went on to define a sixth sense, namely a muscle sense.

SENSORY PHYSIOLOGY

Johannes Müller (1801–1858), a brilliant German physiologist working in Bonn and then Berlin, formulated certain sensory principles in 1826[7] and elaborated on them in his *Handbuch der physiologie der Menschen* (published between 1834 and 1840).[D] These principles came to be known as the doctrine of specific nerve energies (a term with a somewhat obscure meaning). Müller proposed, in essence, that each type of sensory nerve, when stimulated, produces its own specific sensation regardless of the stimulus, because of the part of the brain with which it connects. In fact, Bell had reached a similar conclusion some years earlier, in his 1811 book. Given the biological importance of the concepts, it is worth considering exactly what Bell wrote:

> It is admitted that neither bodies nor the images of bodies enter the brain. It is indeed impossible to believe that colour can be conveyed along a nerve; or the vibration in which we suppose sound to consist can be retained in the brain: but we can conceive, and have reason to believe, that an impression is made upon the organs of the outward senses when we see, or hear, or taste.
>
> In this inquiry it is most essential to observe, that while each organ of sense is provided with a capacity of receiving certain changes to be played upon it, as it were, yet each is utterly incapable of receiving the impressions destined for another organ of sensation.
>
> It is also very remarkable that an impression made on two different nerves of sense, though with the same instrument, will produce two distinct sensations; and the ideas resulting will only have relation to the organ affected. . . .
>
> When the eye-ball is pressed on the side, we perceive various coloured light. Indeed the mere effect of a blow on the head might inform us, that sensation depends on the exercise of the organ affected, not on the impression conveyed to the external organ; for by the vibration caused by the blow, the ears ring, and the eye flashes light, while there is neither light nor sound present.

[D] Translated into English as *Elements of Physiology* (1840–1843).

It may be said, that there is here no proof of the sensation being in the brain more than in the external organ of sense. But when the nerve of a stump is touched, the pain is as if in the amputated extremity...[E].

If light, pressure, galvanism, or electricity produce vision, we must conclude that the idea in the mind is the result of an action excited in the eye or in the brain, not of anything received, though caused by an impression from without.[3]

In other words, Bell stated that sensations are not conducted to the brain as sensations but as something that the brain interprets as sensations. An external stimulus excites a sensory organ, and these organs normally respond only to a very specific stimulus and in a very specific manner. The same stimulus applied to two quite different types of sensory organ or nerve will produce different sensations, such as when a sharp steel point elicits a sensation of touch from stimulation of certain papillae on the tongue and a metallic taste from other papillae. Different stimuli applied separately to the same sensory organ produce the same sensation. Thus, both a visual and a mechanical stimulus to the eye leads to the sensation of light flashes but not to a smell, a sound, and so forth. These sensations must arise in the brain or peripheral sensory organ, but—Bell argued—the former is more likely because of the occurrence of phantom sensations. After limb amputation, for example, stimulation of nerves in the stump produces pains in the missing limb.

Bell in his writings gave many descriptions of phantom pain or other phantom sensory symptoms to support his belief that sensation is experienced in the mind. These perceptions relate to a limb or bodily part that no longer belongs to the body. They occur most commonly following amputation. In an attempt to reduce the risks of developing phantom pains, Bell and others recommended drawing down the nerve and cutting it as short as possible during amputation so that its end retracts into muscle, away from the scar and the pressure of a prosthesis.[8] The entity of phantom pain was to gain increasing recognition after the seminal publications of Silas Weir Mitchell during and after the American Civil War and has come to be associated with Mitchell's name.[F]

[E] He gave another example—but, perhaps out of delicacy, in Latin—of a phantom sensation in patients with destructive lesions of the penis: Quando penis glandem exedat ulcus, et nihil nisi granulation maneat, ad extremam tamen nervi pudicae partem ubi terminator sensus supersunt, et exquisitissima sensuts gratificatio. [When the glans penis is ulcerated and nothing remains but granulation tissue, the most exquisite sensual gratification still remains at the end of the pubic nerves where sensation terminates.]

[F] Silas Weir Mitchell (1829–1914), physician and literary figure, is regarded as the father of American neurology. The now-classic monograph that he coauthored in 1864, *Gunshot Wounds and Other Injuries of Nerves*, provided good descriptions of such phenomena as phantom limb and causalgia (severe burning pain, usually arising after nerve injury). Mitchell was elected the

Clearly, Bell and Müller had similar ideas about many aspects of sensory physiology. Both provided observations to back up their beliefs, but Bell's book initially was circulated only privately, whereas Müller gave the doctrine a public and prestigious face. It is unfortunate, then, but not entirely surprising that Bell has received little public recognition of his contribution.[9,10] Given the acrimonious squabbling with Magendie and Mayo regarding priority—where Bell's claims were on shakier ground—it is ironic that claims for originating the doctrines that now bear the name of Müller did not arise, for there was "no single principle that was new with Müller."[11] Bell—it is true—later commented briefly, as an aside almost, that

> my original experiments . . . have been attributed to foreign physiologists. The ignorance of what has been done in England, may be, for strangers, an excuse for maintaining these opinions as their own; but the authors at home, who should have known what has been taught in this country, are inexcusable when they countenance these assumptions.[12]

The major conclusion articulated by both Bell and Müller was "that we are directly aware, not of objects, but of our nerves themselves; that is to say, the nerves are intermediaries between perceived objects and the mind and thus impose their own characteristics on the mind."[11] Philosophers for years—centuries—have grappled with such concepts, which were not original to either neuroscientist.

THE SIXTH SENSE

The year 1826 was a busy one for Bell, as he continued his work on the nervous system. The University of London (subsequently renamed University College London) was founded in that year, and Bell played a role in its establishment (see Chapter 11). It was also the year that the first Cowes Regatta was held in the Isle of Wight and HMS Beagle sailed on its first voyage; and the year that John Adams, second president of the United States, and Thomas Jefferson, his successor, both died. Others who died that year included René Laennec, the French physician who invented the stethoscope, and Philippe Pinel—founder of modern psychiatry—who unchained the lunatics under his care in Paris.

In a major paper communicated in January 1826 to the Royal Society by its president on his behalf, Bell indicated that muscle receives innervation by both its motor nerve and sensory twigs, and he then set about reasoning why this

---

first president of the American Neurological Association when that organization was established in 1875 and is recalled by the award named after him by the American Academy of Neurology (founded in 1948) to recognize investigative work by those entering the specialty.

should be the case.[13] In considering this paper, the issue concerning to whom the credit is due for distinguishing between the origin of the motor and sensory fibers (discussed in Chapter 7) can be set aside. Bell pointed out that there is a muscle sense because of which

> we feel the effects of over exertion and weariness, and are excruciated by spasms, and feel the irksomeness of continued position. We possess a power of weighing in the hand:— what is this but estimating the muscular force? We are sensible of the most minute changes of muscular exertion, by which we know the position of the body and limbs, when there is no other means of knowledge open to us. If a rope-dancer measures his steps by the eye, yet on the other hand a blind man can balance his body. In standing, walking, and running, every effort of the voluntary power, which gives motion to the body, is directed by a sense of the condition of the muscles, and without this sense we could not regulate their actions.[13]

In other words, without a sense of awareness of the position of the body and limbs, it would not be possible to regulate properly the actions of muscles—in order to move a limb accurately, for example, it is necessary to know where it is in space. Because of this sense, we are able to stand upright with stability. There is, Bell therefore concluded, a *circle of nerves* between the brain and muscle: One nerve is motor, whereas the other is sensory, "giving the sense of the condition of the muscle to the brain." He called this sense the "sixth sense."[G] He is correct, of course, and he can justifiably claim credit for recognizing what today is called the proprioceptive sense.

Although others had previously hinted at such a muscle sense, including Aristotle and Descartes and their followers, they had failed to understand or expand on its physiological implications or importance. Bell himself had given some hint of his thoughts in 1823 when writing on eye movements and the location of visual after-images.[14,H] The sixth sense is now known to be subserved by special sensory endings that exist in the muscles (the muscle spindles) and tendons (Golgi tendon organs) to reflect muscle length and muscle tension, respectively, and their nerve fibers are derived from the posterior (afferent) roots.

A number of cases with impairment of this sense to varying degree have been described in recent years, with the resulting incoordination sometimes being referred to as a sensory ataxia. The extent of motor disruption has varied. One such case—studied in great detail—involved a patient with almost

---

[G] The other five senses (of Aristotle) are vision, hearing, taste, smell, and touch.
[H] He suggested that the direction or location of a visual image or after-image is not determined simply by visual stimulation but also involves information about eye position relating to the activity of the extraocular muscles.

normal power but whose hands were clumsy and relatively useless in daily life due to deafferentation following a severe peripheral neuropathy. He was unable to grasp a pen and write, fasten his shirt buttons, or hold a cup in one hand. His difficulty related in part to the absence of any automatic reflex correction in voluntary movements and to an inability to sustain constant levels of muscle contraction for longer than one or two seconds without visual feedback.[15]

An even more dramatic case was described by Oliver Sacks in *The Man Who Mistook His Wife for a Hat and Other Clinical Tales*.[16] In that book, a chapter is devoted to the tale of a young woman who—while in hospital receiving prophylactic antibiotics prior to gallbladder surgery—developed unsteadiness of gait, clumsiness of the hands, and a sense of being "disembodied," by which she meant that she could not feel her own body. He symptoms were dismissed by a psychiatrist as due to "preoperative anxiety" and "hysteria," but they continued to worsen so that, by the day of surgery, she could not sit up, stand, or hold things in her hands, could not even feed herself because she was so clumsy, and had a slack expressionless face. It eventually became apparent that she had lost all appreciation of muscle and joint position sense with, in addition, some very minor changes in other sensory modalities and in muscle strength, due to an inflammatory process selectively involving certain peripheral nerve fibers. Recovery did not occur, and she was left feeling that she did not own her body—that it had no sense of itself. She eventually learned to move fairly well by using visual feedback to compensate for her peripheral sensory loss, but remained disabled and with a loss of her body image.

The term "circle of nerves," as used by Bell, sounds suggestive of a nervous reflex pathway, but Bell was referring to the dual nerve supply of muscles, not to reflexes and the central mechanisms involved in them.[17] It was Marshall Hall (1790–1857)—abrasive, angry, vituperative even—who elaborated on the reflexes in the first half of the nineteenth century, following the earlier lead of Robert Whytt (1714–1766) and Jiri Prochaska (1749–1820) that showed the spinal cord and other central structures to have an important role in generating the muscle responses to various stimuli.

AND SO TO MOVEMENT

In order to move a limb, it is not sufficient simply to activate the relevant muscles, and Bell was remarkably farsighted in his thinking about what else was required. If it is intended, for example, to bend the thumb, it becomes necessary not only to activate the appropriate flexor muscles but also to inhibit the activity of any opposing (extensor) muscles. Descartes had discussed this in the seventeenth century, but his thoughts were soon forgotten. Bell considered the issue anew. He pointed out in 1823, in just a footnote to a paper on eye movements, that the nerves are usually thought of as stimulating muscles, whereas they also act in the "opposite capacity." This allows one muscle to relax as another contracts. He demonstrated this experimentally:

> I appended a weight to a tendon of an extensor muscle, which gently stretched it and drew out the muscle; and I found that the contraction of the opponent flexor was attended with a descent of the weight, which indicated the relaxation of the extensor. To establish this connection between two classes of muscles, whether they be grouped near together, as in the limbs, or scattered widely as the muscles of respiration, there must be particular and appropriate nerves to form this double bond, to cause them to conspire in relaxation as well as to combine in contraction.[18]

Bell's concept of what is now called reciprocal inhibition was remarkable because, at that time, only positive effects were said to follow nerve stimulation. He seems at least initially to have believed that the nerve acts peripherally at the muscle itself to cause it to relax, however, and this is not correct—the inhibitory effect on limb muscles is exerted in the spinal cord. But within a few years, he came to the realization that the spinal cord must have an integrative action of its own. How else to explain the ability of a chicken to run around with its head cut off? The action of the limbs must be coordinated with each other and with the trunk, and as this coordination occurs even in decapitated animals, it must occur in the spinal cord. The spinal cord is therefore more than simply a passageway for nerve fibers connecting the brain with the periphery. As he stated, this coordination between the body parts, these "combined motions and relations are not established in the brain, the phenomena exhibited on stimulating the nervous system of the decapitated animal sufficiently evince. They must, therefore, depend on an arrangement of fibres somewhere in the spinal marrow."[5] Recognizing the integrative role of the spinal cord—itself an astonishing achievement in the 1830s—he must surely have concluded that this involved an inhibitory spinal function.

Some years after Bell's brief discussion of the topic, the Weber brothers in the 1840s reported that stimulation of the vagus nerve inhibited the heart (arresting its action), and this is indeed a peripheral effect. It was not until 1863 that experimental evidence was obtained by Ivan Setschenov (1829–1905), a Russian physiologist, that inhibition also occurred centrally—limb reflexes, for example, were depressed by stimulating the exposed midbrain. [19] Brown-Séquard in Paris subsequently made much of such central inhibitory effects in trying to understand the operation of the nervous system in health and disease. Charles Sherrington[1] went on to provide clear experimental proof of central

---

[1] Sir Charles Scott Sherrington (1857–1952) was professor of physiology at the University of Liverpool and then at Oxford. His work focused on the function of neurons and their interactions.

inhibitory processes and their importance for the integrative action of the nervous system, work for which he was later honored by a Nobel Prize.[19] Nowadays, reciprocal activation is recognized as occurring in the spinal cord or brainstem such that activation of a muscle leads to the inhibition of its antagonist (opposing) muscles.

THE LONG THORACIC NERVE

In 1821 and 1822, Bell described a new nerve, somewhat similar to the familiar phrenic nerve, which he named the external respiratory nerve.[20,21] It is now known as the long thoracic nerve of Bell. Arising primarily from the fifth and sixth cervical segments, it descends in the neck and passes through the axilla behind the brachial plexus and axillary vessels, to supply the serratus anterior muscle on the outside of the ribs. Bell thought it arose also from the fourth cervical segment and attributed to it a respiratory function in elevating the ribs when the scapula is fixed or pulled back by the trapezius muscle.

The nerve may be damaged by injury or surgery, by pressure from heavy weights (rucksack palsy), and by certain other disorders, leading to weakness of the serratus anterior muscle, manifest by winging of the scapula—pushing the outstretched arm against a wall leads to abnormal protrusion of the shoulder blade away from the rib cage.

REFERENCES

1. Olmsted JMD: The aftermath of Charles Bell's famous "Idea." *Bull Hist Med* 1943; 14: 341–351.
2. Clarke E, Jacyna LS: *Nineteenth Century Origins of Neuroscientific Concepts.* p. 111. University of California Press: Berkeley, 1987.
3. Bell C: *Idea of a New Anatomy of the Brain; Submitted for the Observations of His Friends.* Strahan & Preston, 1811. Reprinted in *Medical Classics* 1936; 1: 105–120.
4. Bell C: *An Exposition of the Natural System of the Nerves of the Human Body.* p. 22. Spottiswoode: London, 1824.
5. Bell C: On the functions of some parts of the brain, and on the relations between the brain and nerves of motion and sensations. *Philos Trans R Soc Lond* 1834; 124: 471–483.
6. Bell C: *An Exposition of the Natural System of the Nerves of the Human Body.* p. 2. Spottiswoode: London, 1824.
7. Müller J: *Zur vergleichenden Physiologie des Gesichtssinnes des Menschen und der Thiere.* Knobloch: Leipzig, 1826.

His book, *The Integrative Action of the Nervous System*, a classic, was based on the Silliman lectures he delivered at Yale University in 1904. He had a major role in showing the presence and importance of reciprocal innervation, by which the activation of a muscle leads to the inhibition of its antagonist (opposing) muscles. He shared the Nobel Prize in Physiology or Medicine in 1932.

8. Crumplin M: *Men of Steel. Surgery in the Napoleonic Wars.* p. 313. Quiller: Wykey, UK, 2007.
9. Finger S, Wade NJ: The neuroscience of Helmholtz and the theories of Johannes Müller. Part 2: Sensation and perception. *J Hist Neurosci* 2002; 11: 234–254.
10. Cassedy S: A history of the concept of the stimulus and the role it played in the neurosciences. *J Hist Neurosci* 2008; 17: 405–432.
11. Boring EG: *A History of Experimental Psychology.* 2nd edition. pp. 81–82. Appleton-Century-Crofts: New York, 1957.
12. Bell C: *The Hand, Its Mechanism and Vital Endowments, as Evincing Design.* p. 151. Pickering: London, 1833.
13. Bell C: On the nervous circle which connects the voluntary muscles with the brain. *Philos Trans R Soc Lond* 1826; 116: 163–173.
14. Bell C: On the motions of the eye, in illustration of the uses of the muscles and nerves of the orbit. Communicated by Sir Humphry Davy, Bart. P. R.S. *Philos Trans R Soc Lond* 1823; 113: 178–179.
15. Rothwell JC, Traub MM, Day BL, Obeso JA, Thomas PK, Marsden CD: Manual motor performance in a deafferented man. *Brain* 1982; 105: 515–542.
16. Sacks O: The disembodied lady. pp. 43–54. In *The Man Who Mistook His Wife for a Hat and Other Clinical Tales.* Harper-Perennial: New York, 1985.
17. Clarke E, Jacyna LS: *Nineteenth Century Origins of Neuroscientific Concepts.* p. 112. University of California Press: Berkeley, 1987.
18. Bell C: Second part of the paper on the nerves of the orbit. Communicated by Sir Humphry Davy, Bart. Pres. R.S. *Philos Trans R Soc Lond* 1823; 113: 289–307.
19. Sherrington CS: Inhibition as a coordinating factor. Nobel Lecture, December 12, 1932. pp. 278–289. In *Nobel Lectures, Including Presentation Speeches and Laureates' Biographies: Physiology or Medicine, 1922–1941.* Amsterdam: Elsevier (for the Nobel Foundation), 1965.
20. Bell C: On the nerves; Giving an account of some experiments on their structure and functions, which lead to a new arrangement of the system. *Philos Trans R Soc Lond* 1821; 111: 398–424.
21. Bell C: On the nerves which associate the muscles of the chest, in the actions of breathing, speaking, and expression. Being a continuation of the paper on the structure and functions of the nerves. *Philos Trans R Soc Lond* 1822; 112: 284–312.

9

CLINICAL OBSERVATIONS ON THE NERVOUS SYSTEM

With his remarkable knowledge of anatomy and his analytical mind, Bell developed into an outstanding clinical neurologist even before the specialty had been invented. Thus it was that in his later years, when finally he returned to the University of Edinburgh as professor of surgery, referrals and requests for consultation were often for him to provide a neurological opinion rather than to perform surgical operations. His clinical observations regarding motor or sensory disturbances involving the face are of particular interest given his interests in the facial expression of emotions (see Chapter 4) and the innervation of the face (see Chapter 7).

LOOKING AT THE FACE

Weakness of one side of the face occurs for many different reasons. Injury or a pathological process involving the nerve to the muscles of facial expression may be responsible, as may involvement of the lower motor neurons in the brainstem from which the nerve originates. In such cases, all facial movements are affected to a greater or lesser extent, depending on the completeness of the lesion, and there is drooping of the face and angle of the mouth, sometimes with inability to close the eye on the affected side. In other instances, a stroke or tumor, for example, occurs centrally and affects the upper motor neurons in a part of the brain that—in turn—connects with those brainstem neurons. The weakness is then more limited, with relative sparing of the forehead, and may affect volitional but not emotive movements, or vice versa.

Peripheral facial palsy was described by physicians of the Greco-Roman era,[1] by the Persian physicians Razi in the ninth century and Avicenna in the eleventh century,[1] by Cornelis Stalpart van der Wiel from The Hague in the seventeenth century,[2] and by Nikolaus Anton Friedreich from Würzburg in the eighteenth century.[3,4] Friedreich described three patients with the abrupt onset of unilateral facial paralysis that cleared completely over the subsequent weeks and months. Facial sensation was intact, and there was no motor or sensory abnormality elsewhere. He differentiated the disorder from a central cause by the lack of other associated symptoms, and he attributed it to a "rheumatic" cause because of a history of exposure to cold and of fever, chills, and pain and swelling about the ear. It is not known whether Bell ever read Friedreich's account, which was published in the *Annals of Medicine* in Edinburgh in 1800. In any event, he described a case of his own in 1821, a man whose facial nerve was injured by a local infection, and noted in the same article that partial facial paralysis ("the blight") was not uncommon in young people:

> A man had the trunk of the respiratory nerve of the face injured by a suppuration, which took place anterior to the ear, and through which the nerve passed in its course to the face. It was observed that, in smiling and laughing, his mouth was drawn in a very remarkable manner to the opposite side. The attempt to whistle was attended with a ludicrous distortion of the lips; when he took snuff and sneezed, the side where the suppuration had affected the nerve remained placid, while the opposite side exhibited the usual distortion.[5]

He subsequently described other cases, and the occurrence of peripheral facial palsy of unknown (probably viral) cause is now named after him (Bell's palsy). The lack of any disturbance of facial sensation in these cases confirmed his experimental observations—and those of Herbert Mayo—that the seventh and fifth cranial nerves have different functions (see Chapter 7). It has been suggested that Bell may himself have had a right-sided facial palsy, based on a slight asymmetry of his face in certain portraits,[6] but this is doubtful,[7,8] and in any event it is not the reason that facial paralysis is named for him.

Bell had reported in 1823 "a motion of the eye-ball, which, from its rapidity, has escaped observation. At the instant in which the eye-lids are closed, the eye-ball makes a movement which raises the cornea under the upper eye-lid."[9,10] He considered this to be a normal protective maneuver against injury. When a severe peripheral facial palsy prevents the eye from closing, a visible upward and outward turning of the globe occurs with attempted eye closure (Bell's phenomenon or sign). He demonstrated this in a clinical lecture at the Middlesex Hospital involving 70-year-old Daniel Quick, a street-sweeper who had been tossed by a bull twelve years earlier—the bull's horn went in behind the upright portion of his jaw, tearing the facial nerve, and he was left with severe facial weakness.[11] The patient's eye would roll up involuntarily and without his knowledge on attempted eye closure and when he was asleep.

The clinical features of a severe peripheral facial palsy are now well known. On the affected side, the skin of the forehead is flat, smooth, and unwrinkled. On frowning, the affected eyebrow barely moves except to be pulled toward the opposite side due to the action of the muscles there. The eye cannot be closed either voluntarily or during blinking; on attempted eyelid closure, the eye rolls up. The cheek is flaccid and without wrinkles; on speaking, it puffs out and then collapses as air escapes at the angle of the mouth. The affected nostril is not distended during a sniff or with deep inspiration: "It is a waste of time to put snuff into this nostril: . . . It does not go high enough: . . . [The patient] cannot make it mount." Saliva may dribble from the mouth, which is pulled to the unaffected side. The face becomes distorted during laughter—"at each cachinnation" the affected cheek is puffed out, "flapping like a loose sail" while on the other side the mouth is drawn up, the cheek is wrinkled, and the eye-lids puckered.[12] Bell's description was very detailed, and this disorder, more than any other, is now associated with his name. Fortunately, in many patients, Bell's palsy recovers completely without treatment, and most patients are satisfied with the final outcome even if slight weakness persists. The differences between a peripheral and central facial palsy, discussed previously, were illustrated by Bell in various case histories in a monograph on the human nervous system that he published in 1830 and that went through several editions.[12] The book contained papers that had previously been read before the Royal Society and published in its *Philosophical Transactions*, edited as discussed previously, as well as an appendix with neurological case histories and observations.

Some have wondered whether the enigmatic smile of the Mona Lisa (la Giaconda) in Leonardo Da Vinci's painting is the result of Bell's palsy. Certainly her face is slightly asymmetrical, but whether this is normal, due to a facial weakness, or the result of a contracture (which sometimes follows a Bell's palsy) cannot be concluded.

The facial nerve can also be affected in other ways. Patients sometimes develop an involuntary irregular twitching or more vigorous contraction of one side of the face, which may be so sustained that the face briefly goes into spasm. These hemifacial spasms may begin months or years after a Bell's palsy or—more commonly—in patients who have never had facial weakness. They have been named after Bell and are also known as convulsive tics. They typically begin about the eyelids and then spread down to the cheek and side of the mouth.[13] They are now attributed to compression of the facial nerve near the brainstem by a small adjacent blood vessel. Modern treatment usually consists of injections of botulinum toxin into the facial muscles to prevent the involuntary movements or a surgical procedure (microvascular decompression) to relieve pressure on the nerve.

THE NUMB CHIN

One of the main reasons to publish details of individual cases is to allow others to recognize similarities with cases of their own that might otherwise be disregarded and thereby facilitate the recognition of new syndromes or diseases.

Who, for example, would have suspected the sinister implications of a numb chin? One of the very last cases that Bell described in his 1830 monograph on the nervous system was of an "elderly maiden lady" with breast cancer who consulted him for "an insensibility of the lower lip: her attention was called to this by feeling only one half of the cup in drinking." Bell found that she could not feel touch on the left lower lip and that she had a hard tumorous mass affecting the jaw on that side, presumably pressing on the inferior alveolar nerve (a branch of a division of the fifth cranial nerve).[14] This nerve exits from a bony canal on the side of the chin, dividing into sensory branches to the lip and chin. In the past fifty years, isolated numbness in this region has been reported in a number of cases of metastases to the lower jaw or skull base from different cancers and sometimes is the first manifestation of such malignancy.[15,16] The so-called numb chin syndrome—first described by Bell—has thus gained an important but ominous significance.

MOTOR NEURON DISEASES

Bell described a number of patients with neuromuscular diseases, including some with diseases that probably had never been reported previously. However, he failed to identify these disorders as distinctive or to collect a sufficient number of cases to show their common features, and he did not set them off from other disorders that they might have resembled. In consequence, their eponymous designation has gone to later authors who recognized their individuality rather than to Bell, who described them without realizing their true significance. It is only on looking back at these cases that they assume some historical interest, although the diagnosis is not always secure.

Progressive spinal muscular atrophy is a disease with insidious onset and gradual progression—motor neurons in the spinal cord die, so the muscles they supply become progressively weaker and wasted. The first recorded case is probably that of Bell (case CLXXIX), who described a patient with progressive muscular wasting and weakness, first in the upper extremities and then in the legs, which began when he was thirteen years old. The boy had no pain or cognitive disturbance and was otherwise well. Bell concluded that the disorder was neurogenic rather than myogenic, and he correctly suggested that a central process was likely.[17]

If the upper motor neurons (in the brain) or their connection to the spinal cord (the corticospinal tract) are involved as well as the spinal (lower) motor neurons, there is typically some combination of weakness, wasting, stiffness, and spontaneous flickering affecting the limb and bulbar muscles diffusely, and the muscle stretch reflexes may be increased or diminished. The disorder is now known as amyotrophic lateral sclerosis (ALS), often called Lou Gehrig disease in the United States after the baseball player who died of the disease in 1941. Bell provided the first description of such a patient, a woman—"upwards of fifty years of age, of a delicate, nervous temperament"—with a progressive motor disorder that came on a few weeks after a fall and affected the limbs and tongue. She eventually developed severe weakness of the upper and lower limbs, twitching of the muscles, and a bulbar palsy with impairment of speech and swallowing, but she had no sensory abnormalities (although she did experience some numbness and paresthesias in the fingers) and no cognitive or sphincter disturbance. For a while, she was able to communicate with others by directing a small stick to letters printed on pasteboard, but for some weeks before her death she was unable to do even that. Autopsy showed degenerative changes in the anterior part of the spinal cord.[18]

Finally, Bell reproduces in his monograph a referral letter from a Dr. R. W. Robinson of Preston written in 1825 about a seventy-year-old woman in otherwise good health who developed increasing difficulty in moving her tongue, slurring of speech that progressed to anarthria (mutism), drooling of saliva, and increasing difficulty in swallowing, with occasional choking. "The tongue itself is soft and pulpy. . . . She cannot hack up any thing from the throat, nor draw any thing from the posterior nares by a back draught." Facial sensation and power were normal, as were her limbs. Bell attributed her symptoms to disorder of the hypoglossal or twelfth cranial nerve[A]—that is, to a disorder of the motor nerve of the tongue[19]—but it was probably more extensive than that. This is the first known example of progressive bulbar palsy,[20,21] a disorder that may occur for a variety of reasons including ALS. The "bulb" in this context refers to the bulb-shaped lower brainstem, which contains the nuclei of the ninth to twelfth cranial nerves supplying the muscles of speech, swallowing, and the tongue. A bulbar palsy thus involves weakness of these muscles, but there may be more extensive findings to suggest the nature of the underlying lesion. In Robinson's patient, without more clinical details a definite etiological diagnosis is difficult to establish.

MUSCULAR DYSTROPHIES

There are certain disorders in which muscle wasting and weakness result from primary involvement of the muscles. Bell gave the first description of what may well have been a muscular dystrophy.[22-24] The patient was a young man who had progressive weakness and wasting of the lower extremities, beginning about

---

[A] Bell called the responsible nerve the ninth cranial nerve according to the system in use at the time, as noted in footnote H, Chapter 7.

the thighs when he was ten years old, so that he had difficulty in arising from a seated position. "It is now curious to observe," wrote Bell, "how he will twist and jerk his body to throw himself upright from his seat." He had a slight lumbar kyphoscoliosis but no involvement of the upper limbs and no sensory or sphincter disturbance. Again, the diagnosis remains uncertain in the absence of pathological confirmation, but the clinical description strongly suggests a muscular dystrophy.

MYOTONIA

In 1836, Bell described a particular "affection of the voluntary nerves":

> The most common instance of this is an impediment of speech, when the consent of the muscles is imperfect; but this sometimes extends to all the voluntary muscles of the body. I find that some are capable of lifting a heavy weight, or walking fifteen or twenty miles, and yet they have not the proper command of their limbs: There is an insecurity and want of confidence in the motions of the body, which overtakes them upon any excitement; a paralysis of the knees which prevents the individual from putting one leg before the other, and which endangers their falling. Thus a gentleman, capable of great bodily exertion, on going to hand a lady to the dining-room, will stagger like a drunken man; and in the streets any sudden noise, or occasion of getting quickly out of the way, will cause him to fall down, and in this manner a want of confidence produces a nervous excitement which increases the evil. With confidence the power of volition acts sufficiently; there is neither defect of speech nor irresolution in the motion of the limbs when the person is at ease or under a flow of spirits.
>
> Such cases are very curious in their details, as exhibiting an extraordinary degree of incapacity for the affairs of life proceeding from slight defects. There is neither disease of mind nor of bodily organs; the corporeal frame is perfect; the nerves and muscles are capable of their functions and proper adjustments; the defect is in the imperfect exercise of the will, or in that secondary influence which the brain has over the relations established in the body.[25]

Many authors have considered this case to be the first recorded one of myotonia congenita,[26,27] but the description is really too vague to allow a confident diagnosis.

Nevertheless, Julius Thomsen (1815–1896), a Danish physician, was able to recognize his own illness and that of several family members from Bell's description. Whether this encouraged him to write his own account is unclear, but he published it only to defend his son when the young man was accused of malingering to avoid military service.[28] In any event, he subsequently reported all the cardinal clinical features of the disease,[29,30] but he credited Bell with having first described the malady. Thomsen thus both described and suffered from the disorder that now bears his name. As Bell failed to give the disease a name or to distinguish it from other disorders, his contribution is not widely recognized. Thomsen's disease, which has a dominant mode of inheritance, is characterized by muscle stiffness and cramps that begin in childhood, are enhanced by cold or inactivity, and are relieved by exercise. There is delayed relaxation of a contracted muscle. Falls may occur from hasty movements or an inability to rapidly correct a loss of balance. The muscles may be enlarged; they are not weak. A recessively inherited form associated with mild weakness and atrophy of distal muscles has also been recognized.

MOVEMENT DISORDERS

Movement disorders are characterized by excessive, inappropriate, or diminished movement in the absence of weakness or sensory disturbance. They are often associated with pathological changes in the basal ganglia, which lie within the depths of the cerebral hemispheres. Some abnormal movements—the dystonias—are so slow and sustained that they are almost abnormal postures. Dystonic posturing is commonly exacerbated or brought on by voluntary activity, sometimes occupational.

Writer's cramp ("scrivener's palsy") was well recognized in the nineteenth century. Bell may have given its earliest description in 1830: "I have found the action necessary for writing gone, or the motions so irregular as to make the letters be written zigzag, whilst the power of strongly moving the arm, or fencing, remained."[31] He subsequently noted other occupational cramps:

> Here is a musician who, being a fiddler, has been obliged to change his instrument, because the extensor muscles drew the bow off the strings of the violin. He can steady his arm to play upon the harp, by pressing his elbow to his side.[32]

Another patient could not use a paper-folder, the motion of the wrist being affected, even though examination revealed no weakness.[33] Such cramps are task-specific disorders characterized by impairment of certain occupationally related movements. Bell's accounts of them were often reduced to one or two lines without any detailed description, and consequently there are few references to them by modern authors.

Bell also described several patients with cervical dystonia (torticollis), the head being tilted or twisted to one side.[34] In some cases, the disorder was progressive, painful, worsened with stress, accompanied by a head tremor, and associated with hypertrophy of the neck muscles. Some patients had developed sensory tricks to relieve the abnormal posturing, such as light touch or pressure on the face or back of the head.³⁵ Treatment, including restraint with a soft collar or steel apparatus, was generally unhelpful. Bell believed that torticollis is an organic disorder.ᴮ He correctly emphasized that it is not a primary disease of muscle—as some believed—but a disease of the nervous system.

He attributed it to a disorder of the nerves to the neck muscles. For example, one of his patients at the Middlesex Hospital had torticollis that he attributed to a disorder of the accessory nerve (the eleventh cranial nerve), accompanied by pain down the inner side of the left arm that was apparently "relieved by the application of leeches to the temples and a blister to the nape of the neck."³⁶ The *Lancet*—just two years old—in describing the patient in its Hospital Reports, disagreed with Bell and held that the motor abnormality was more extensive. "We have frequently had occasion to notice," the anonymous writer went on to state, "the *very ingenious manner* [sic] in which Mr. Bell perverts facts, in order to meet his own particular views of a case." Discourtesy and invective was part of that journal's style under the editorship of its radical founder, Thomas Wakley (see Fig. 6-1), whose efforts at social and health care reform were hampered by unnecessary rudeness directed at the medical establishment. Torticollis and other dystonic disorders that develop without a specific local cause are now thought to originate in the brain, in the basal ganglia. Modern treatment is aimed at reducing the excessive, inappropriate, and often painful muscle contractions that result by injecting botulinum toxin into the involved muscles. This reduces the effectiveness of the nerve impulses that trigger the muscle contractions and thereby provides temporary relief. It does not, of course, cure the underlying neurological problem.

ATLANTOAXIAL DISLOCATION

Bell in 1830 reported a patient with pharyngeal ulceration (probably syphilitic) who died suddenly from dislocation of the atlas from the second vertebra of the neck. At autopsy, it was found that "the ulcer had destroyed the transverse ligament, which holds the process of the dentata in its place. In consequence of the failure of this support, the process was thrown back, so as to compress the spinal marrow."³⁷ This was a remarkable case, the first of its kind to be reported. The condition has since been named for the French physician Pierre Grisel, who— one hundred years later—reported children with pharyngitis and atlantoaxial dislocation related to an infectious process. Patients typically present with

---

ᴮ A number of later clinicians, including Charcot in Paris, concluded that torticollis did not have a structural basis or that it was a psychogenic disorder or "névrose", and it is only in the past fifty years that an organic basis has come to be accepted widely. Further details are provided in

painful torticollis and aversion to head rotation. The condition can be fatal if untreated.

A number of cases of post-traumatic atlantoaxial dislocation leading to unexpected death can be found in Bell's monograph. The risk of atlantoaxial dislocation is one of the reasons for immobilizing the head after trauma to the head or neck until imaging studies reveal the extent of injury.

TIC DOULOUREUX AND REFERRED PAIN

Johann Laurentius Bausch (1605–1665), one of the founders of the Imperial Leopoldina Academy of Natural Sciences and its first president,[C] suffered with tic douloureux, as described in his eulogy published in 1671.[38] This early report was supplemented when John Locke (1632–1704), the English physician-philosopher, provided a full clinical account of the disorder. His patient was the Countess of Northumberland, wife of the British ambassador in France, whom he saw in Paris in December 1677 with a recurrence of tormenting facial pain. Some months previously, she had suffered similar pain, which was not helped when her French physicians extracted two teeth from the affected side. Fortunately, her symptoms resolved spontaneously, for there was nothing effective that the English physician could do, although he consulted with several colleagues back home and treated her with purgatives.[39] The designation *tic douloureux* was used by Nicolas André in 1756 to refer to the facial contortions and grimaces that may be associated with such extreme facial pain and that he attributed to nerve compression.[40] The disorder was described further and in detail by John Fothergill in 1773.[D]

Bell saw many patients with tic douloureux and elaborated on its features and possible causes. As the functions of the fifth cranial nerve became clearer and were distinguished from those of the seventh nerve (see Chapter 7), the disorder came to be referred to as *trigeminal neuralgia*. Stabs of pain—like flashes of lightning—occur in the territory of a branch of the trigeminal nerve, either spontaneously or after minor stimuli (such as light touch, eating, or exposure to a draft) from which the patient tries to protect the face. Patients have been

---

Munts AG, Koehler PJ: How psychogenic is dystonia? Views from past to present. *Brain* 2010; 133: 1552–1564.

[C] The Imperial Leopoldina Academy of Natural Sciences, founded in 1652, is perhaps the oldest academy of science in the world. It became the National Academy of Sciences in Germany in 2008.

[D] John Fothergill (1712–1780) was a London physician, botanist, philanthropist, and active member of the Society of Friends. His botanical collection in his Essex gardens was reputedly second only to that in Kew Gardens among all the collections in Europe. He was a friend of Benjamin Franklin, and together they attempted unsuccessfully to negotiate an agreement between the colonies and Britain to prevent the American War of Independence. On his death, Franklin remarked that a worthier man had never lived.

known to die from malnutrition because eating triggers the pain. Bell's clinical approach resembles that of the modern neurologist. The pain is to be distinguished by clinical examination from that caused by dental or gum disease (which responds to local treatment) and from tumor involvement or intrinsic disease of the nerve. Dental extractions and local nerve blocks provide no lasting benefit for tic douloureux. Nerve pain related to intrinsic disease or a tumor is more constant and accompanied by facial numbness and sensory loss that do not occur in trigeminal neuralgia. Bell noted that cutting either the seventh nerve or branches of the fifth nerve is without benefit and warned against these surgical procedures:

> The division of the branches of the fifth, though it has been practiced by every surgeon of eminence during the last half century, is not to be done.... The advantage is temporary, if any, and the root of the evil is not reached.[41]

He related the disorder to deranged bowel function and reported success with purgatives.[42] This treatment, which now seems curious, was neither novel nor eccentric at the time; it had been widely used for many years. The tendency of the disorder to remit and relapse spontaneously may have helped to foster the illusion of benefit.

It is only since the 1960s that certain medications (antiseizure drugs) have been found to help. Modern surgical treatment is by "microvascular decompression." This is based on the belief that compression of the nerve close to its roots, typically by a small blood vessel, leads to regional demyelination, which in turn may cause pain relating to hyperexcitability of the trigeminal nerve nucleus; spread of electrical impulses from demyelinated fibers mediating light touch to adjacent pain fibers may account for the triggering of painful spasms by various tactile stimuli.

The pain from a bad tooth may be felt diffusely in the face. This led Bell, in an essay on tic douloureux, to discuss referred pain—that is, pain experienced in a site other than where it arises. Pain from internal organs may be felt in a remote region of the body—for example, cardiac pain may be felt in the chest or left arm; liver disease in the shoulder; and renal colic in the testicle, groin, and front of the thigh. Tenderness or hypersensitvity (hyperalgesia) of the skin, subcutaneous tissues, and muscle may accompany referred pain.

The mechanisms underlying referred pain are obscure. John Hunter's concept of "sympathy" was dismissed by Bell as meaningless and unhelpful. He wondered whether the sensory impressions are somehow reflected to a distant site from their site of origin, but he reasoned that this must be incorrect because sensory input is carried toward the brain, not away from it. If two nerve fibers are bound together and one is carrying pain impulses centrally, he wondered whether it could affect the other fiber, or whether impulses being transmitted centrally by fibers from an internal organ could be transmitted to the more sensitive cutaneous sensory nerve fibers accompanying them. Bell considered various types of referred pain, especially in relation to gastrointestinal disease, and finally concluded that irritation affecting the internal branch of a nerve is felt or attributed to the external branch of the same nerve.[43]

Bell's musings on the basis of referred pain are surprisingly sophisticated. Many years after his death, Henry Head[E] and others showed that disease of different internal organs does indeed cause referred pain to specific areas of the skin and subjacent tissues. With severe or chronic visceral pain, the referred pain becomes more diffuse. Head's explanation, now widely accepted, is as follows:

> A painful stimulus to an internal organ is conducted to that segment of the cord from which its sensory nerves are given off. There it comes into close connection with the fibres for painful sensation from the surface of the body which also arose from the same segment. But the sensory and localising power of the surface of the body is enormously in excess of that of the viscera, and thus by what may be called a psychical error of judgment the diffusion area is accepted by consciousness, and the pain is referred on to the surface of the body instead of on to the organ actually affected.[44]

In other words, the pain is referred to the skin and muscle supplied by nerves from the same spinal segment as the involved organ. In the more general terms of Head, when

> a painful stimulus is applied to a part of low sensibility in close central connection with a part of much greater sensibility the pain produced is felt in the part of higher sensibility rather than in the part of lower sensibility to which the stimulus was actually applied.[44]

Head's explanation is not far removed from the speculation of Bell more than fifty years earlier.

The precise pathophysiology of referred pain is still unknown. The number of primary sensory fibers that branch to innervate both viscera and skin or deeper tissues is limited so that peripheral branching of nerve fibers is unlikely to explain referred pain. However, both visceral and somatic afferent fibers are known to converge on the same sensory neurons in the spinal cord and more centrally. Thus, some sensory neurons in the spinal cord are activated by noxious stimuli from both visceral structures and skin or muscle. A misinterpretation of

---

[E] Henry Head (1861–1940) was an English neurologist and polymath who undertook experimental work on sensation (including a study on the effects of severing nerves in his own arm), the segmental innervation of the skin, spinal cord injuries, and speech. He was knighted in 1927.

such sensory input by more central structures involved in pain perception may therefore lead to the referral of visceral pain to other parts of the body (perhaps based on previous experience and the more common origin of somatic pain). Central sensitization of nerve cells in the spinal cord by continuing or increasing pain in a visceral structure may also occur, leading to earlier firing of neurons or a greater number of neurons being activated.[45]

REFERENCES

1. Pearce JMS: Early observations on facial palsy. *J Hist Neurosci* 2015; 24: 319–325.
2. van de Graaf RC, Nicolai JP: Bell's palsy before Bell: Cornelis Stalpart van der Wiel's observation of Bell's palsy in 1683. *Otol Neurotol* 2005; 26: 1235–1238.
3. Bird TD: Nicolaus A. Friedreich's description of peripheral facial nerve paralysis in 1798. *J Neurol Neurosurg Psychiatry* 1979; 42: 56–58.
4. Pearce JMS: Bell's or Friedreich's palsy. *J Neurol Neurosurg Psychiatry* 1999; 67: 732.
5. Bell C: On the nerves; Giving an account of some experiments on their structure and functions, which lead to a new arrangement of the system. *Philos Trans R Soc Lond* 1821; 111: 398–424.
6. Lal R, Weber SAT: About the right facial palsy of Charles Bell. *Arq Neuropsiquiatr* 2009; 67: 783–784.
7. Korteweg SFS, van de Graaf RC, Werker PMN: About the right facial palsy of Charles Bell: Was Sir Charles Bell himself really affected by facial paralysis? Comment on 'Peripheral facial palsy in the past. Contributions from Avicenna, Nicolaus Friedreich and Charles Bell.' *Arq Neuropsiquiatr* 2009; 67: 783.
8. Korteweg SFS, van de Graaf RC, Werker PMN: Sir Charles Bell was not affected by facial paralysis himself! *Arq Neuropsiquiatr* 2010; 68: 321–322.
9. Bell C: Second part of the paper on the nerves of the orbit. Communicated by Sir Humphry Davy, Bart. Pres. R.S. *Philos Trans R Soc Lond* 1823; 113: 289–307.
10. Bell C: On the motions of the eye, in illustration of the uses of the muscles and nerves of the orbit. *Philos Trans R Soc Lond* 1823; 113: 166–186.
11. Bell C: Clinical lecture on partial paralysis of the face, delivered by Mr. Bell, at the Middlesex Hospital. *Lond Med Gaz* 1828; 747–750 and 769–770. Also No. III, pp. vii–xiv. In *The Nervous System of the Human Body*. Longman, Rees, Orme, Brown, and Green: London, 1830.
12. Bell C: *The Nervous System of the Human Body*. Longman, Rees, Orme, Brown, and Green: London, 1830; 3rd edition. Black: Edinburgh, 1836.
13. Bell C: Nos. XII and XIII. pp. 256–258. Also No. CLVII, p. 413. In *The Nervous System of the Human Body*. 3rd edition. Black: Edinburgh, 1836.
14. Bell C: Dec 26. pp. clxiv–clxv. In *The Nervous System of the Human Body*. Longman, Rees, Orme, Brown, and Green: London, 1830.
15. Calverley JR, Mohnac AM: Syndrome of the numb chin. *Arch Intern Med* 1963; 112: 819–821.
16. Horton J, Means ED, Cunningham TJ, Olson KB: The numb chin in breast cancer. *J Neurol Neurosurg Psychiatry* 1973; 36: 211–216.

17. Bell C: No. CLXXIX. pp. 430–431. In *The Nervous System of the Human Body*. 3rd edition. Black: Edinburgh, 1836.
18. Bell C: No. LXVIII. pp. cxxxiii–cxxxv. In *The Nervous System of the Human Body*. Longman, Rees, Orme, Brown, and Green: London, 1830.
19. Bell C: No. CXXII: Affections of the tongue and mouth, &c. pp. 390–391. In *The Nervous System of the Human Body*. 3rd edition. Black: Edinburgh, 1836.
20. Chancellor AM, Mitchell JD, Swingler RJ: The first description of idiopathic progressive bulbar palsy. *J Neurol Neurosurg Psychiatry* 1993; 56: 1270.
21. Mitchell JD: Disorders of anterior horn cells. pp. 347–365. In Critchley E, Eisen A (eds): *Spinal Cord Disease: Basic Science, Diagnosis and Management*. Springer-Verlag: London, 1997.
22. Gardner-Thorpe C: Charles Bell (1774–1842) and an early case of muscular dystrophy: The Third Meryon Society Lecture read at Worcester College, Oxford on 28 July, 2000. *Neuromuscul Disord* 2002; 12: 318–321.
23. Bell C: No. CLXXX: Case of partial paralysis of the lower extremities. pp. 431–432. In *The Nervous System of the Human Body*. 3rd edition. Black: Edinburgh, 1836.
24. Bell C: No. LXXXIX: Case of partial paralysis of the lower extremities. p. clxiii. In *The Nervous System of the Human Body*. Longman, Rees, Orme, Brown, and Green: London, 1830.
25. Bell C: No. CLXXXIV: Affection of the voluntary nerves. p. 436. In *The Nervous System of the Human Body*. 3rd edition. Black: Edinburgh, 1836.
26. Caughey JE: Relationship of dystrophia myotonica (myotonic dystrophy) and myotonia congenita (Thomsen's disease). *Neurology* 1958; 8: 469–476.
27. Lanska MJ, Lanska DJ, Remler B: Vignette. *J Child Neurol* 1990; 5: 316–317.
28. Lanska MJ, Lanska DJ, Remler B: Julius Thomsen. pp. 364–368. In Ashwal S (ed): *The Founders of Child Neurology*. Norman: San Francisco, 1990.
29. Thomsen J: Tonische Krämpfe in willkürlich beweglichen Muskeln in Folge von ererbter psychischer Disposition (Ataxia muscularis?). *Arch Psychiatr Nervenkr* 1876; 6: 702–718.
30. Thomsen J: Nachträgliche Bemerkungen über Myotonia congenita (Strümpell), Thomsen'sche Krankheit (Westphal). *Arch Psychiatr Nervenkr* 1892; 24: 918–923.
31. Bell C: Partial paralysis of the muscles of the extremities. pp. clx–clxi. In *The Nervous System of the Human Body*. Longman, Rees, Orme, Brown, and Green: London, 1830.
32. Bell C: No. CLXXXV: Affection of the voluntary nerves. p. 436. In *The Nervous System of the Human Body*. 3rd edition. Black: Edinburgh, 1836.
33. Bell C: No. CLXXXVI. pp. 436–437. In *The Nervous System of the Human Body*. 3rd edition. Black: Edinburgh, 1836.
34. Bell C: Nos. LXXIV, LXXVI, LXXIX–LXXIV. pp. cxli–clv. In *The Nervous System of the Human Body*. Longman, Rees, Orme, Brown, and Green: London, 1830.
35. Gonzalez-Alegre P: Descriptions of cervical dystonia by Sir Charles Bell. *Mov Disord* 2010; 25: 257–259.
36. Anonymous: Middlesex Hospital. *Lancet* 1825; 4: 189–190.
37. Bell C: No. LXIV. p. cxxvii. In *The Nervous System of the Human Body*. Longman, Rees, Orme, Brown, and Green: London, 1830.

38. Lewy FH: The first authentic case of major trigeminal neuralgia and some comments on the history of this disease. *Ann Med Hist* 1938; 10: 247–250.
39. Dewhurst K: A symposium on trigeminal neuralgia with contributions by Locke, Sydenham, and other eminent seventeenth century physicians. *J Hist Med Allied Sci* 1957; 12: 21–36.
40. Brown JA, Coursaget C, Preul MC, Sangvai D: Mercury water and cauterizing stones: Nicolas André and tic douloureux. *J Neurosurg* 1999; 90: 977–981.
41. Bell C: Tic douloureux. p. 359. In *The Nervous System of the Human Body*. 3rd edition. Black: Edinburgh, 1836.
42. Bell C: Tic douloureux. Nos. LXXVII–LXXXIV. pp. 355–361. In *The Nervous System of the Human Body*. 3rd edition. Black: Edinburgh, 1836.
43. Bell C: Essay IV. On the action of purgative medications on the different portions of the intestinal canal, with a view to remove nervous affections and tic douloureux. pp. 83–104. In *Practical Essays*. Maclachlan, Stewart: Edinburgh, 1841.
44. Head H: On disturbances of sensation with especial reference to the pain of visceral disease. *Brain* 1893; 16: 1–133.
45. Giamberardino MA, Affaitati G, Constantini R: Referred pain from internal organs. *Handb Clin Neurol* 81: 343–361, 2006.

10

FOR GOD AND COUNTRY

Charles Bell was a religious man. This is hardly surprising given that his father and paternal grandfather were clergymen and that his mother was the daughter of a clergyman and granddaughter of a bishop. The woman he had married also came from an ecclesiastical background. Bell reconciled his scientific determinism with his Christian values by means of his beliefs in intelligent (creative) design. As far back as 1806, he stated his views very clearly in his *Essays on the Anatomy of Expression in Painting*, discussed in Chapter 4. Indeed, it was the expression of these views that led Charles Darwin—sixty-six years later—to write his own related work, *The Expression of the Emotions in Man and Animals*, in which he paid homage to and built upon Bell's scientific achievements while objecting to his creationist viewpoint.

In 1819, in his *Essay on the Forces Which Circulate the Blood*, Bell pointed out that humans would be insignificant compared to the grandeur of the universe but for their own wondrous construction. Hydraulic laws could not entirely explain the motion of the blood in living subjects, which also required—he believed—an "attraction as incomprehensible and wonderful as that which retains the planets in their orbits."[1] In contrast to conventional views, he had difficulty in accepting that the circulation was due exclusively to the heart, claiming that arterial contraction helped to move the blood along, under the influence especially of the destination organs or regions. As a consequence, increased activity led to greater local blood flow. He seems in this way to have sensed the existence of the vasomotor reflexes that are now known to exist and that direct regional blood flow based on requirement. He believed that the tortuosity of certain arteries and other vagaries of the circulation were not simply accidents of nature but must be ascribed to the wisdom of the creator in, for example, permitting greater engorgement with blood during activity or for otherwise allowing for alterations in regional blood flow.

In his lectures also, whether at the Great Windmill Street School,[2] the godless University of London (see Chapter 11), or elsewhere, he referred constantly to the elegance of anatomical design, relating this to the wisdom of an all-knowing creator.

Works on intelligent design were common in the eighteenth and nineteenth centuries, relaying concepts that seemed persuasive and were supported by the powerful views of Georges Cuvier (1769–1832), the illustrious naturalist in Paris who held that successive creations occurred after natural catastrophes and opposed any theories of evolution, such as those of Lamarck (1744–1829). Many of the most successful surgeon–anatomists of the day—establishment medical figures—held creationist views, including men such as Astley Cooper, Anthony Carlisle, and John Abernethy, as well as Bell. This provided Wakley and his men at the *Lancet* with the opportunity for ridiculing them at will—an opportunity that they did not waste.[3] Indeed, in a scathing review of the opening address of the winter session by Joseph Henry Green (see Chapter 4), professor of surgery at the newly established King's College London, which touched on creative design, the anonymous reviewer goes on to parody Bell and his Scottish accent: "He never touches a phalanx, and its flexor tendon, without exclaiming, with uplifted eye, and most reverentially-contracted mouth, "Gintilmin, behold the wonderful evidence of *desin!*"[4]

Publications on the topic reached their zenith with William Paley's *Natural Theology*, first appearing in 1802 but going into many editions.[5] Paley (1743–1805), an English cleric and philosopher, argued that the existence of God is shown by the rational design of the creation and does not require divine revelation. He famously regarded the relationship between a watch and a watchmaker as similar to that between the natural world and its creator. Inspection of the innards of a working watch suggests that its several parts were "put together for a purpose" by its creator or designer. In the same way, Paley argued, the complex structure of living organisms and their marvelous adaptations to different environments implies the existence of an intelligent designer. Paley believed that this general approach applied to nature as a whole. The first edition of his book contained thirty-six lithographic plates that illustrated his beliefs, starting with the watch and then going through various anatomical structures.[6]

In 1827, Bell published his *Animal Mechanics: Or, Proofs of Design in the Animal Frame: The Perfection of Design in the Bones of the Head, Spine, and Chest, Shown by Comparison with Architectural and Mechanical Contrivances*. The subtitle of the volume revealed the perspective of the work. It was written at the request of Henry Brougham, by then a rising politician, and was a remarkable success.[7] Probably because of the strengths of his beliefs, his professional standing, and their personal friendship, Bell was then asked to join Brougham in illustrating and editing later editions of Paley's *Natural Theology*,[7] which he did.[8] His notes and appendix, which relate primarily to anatomy and animal physiology, are comprehensive but written in an easy style for a lay readership.

Animal Mechanics, as suggested by its subtitle, was specifically oriented toward intelligent design and was praised by Brougham as "the most original exposition of Paleyism to date."⁹ In his essay, Bell compared the design of the human body with things of human invention—for example, the head to different architectural structures such as the walls and roof of a house, an arch, a bridge, and a dome, and the spine to the mast of a ship, in each case showing the several mechanical advantages possessed by the biological structures. He continued with the chest, the bones and joints of the extremities, and other structures, and he even compared the tendons to cordage. As he developed the concept, his increasing surprise by the magnificent way that form was adapted to function was matched by dismay that the study of anatomy had received so little attention:

> We crowd to see a piece of machinery or a new engine, but neglect to raise the covering which would display in the body the most striking proofs of design, surpassing all art in simplicity and effectiveness, and without anything useless or superfluous.[10]

Knowledge of the animal body must "compel us to acknowledge an Almighty Power in the creation," he concluded.[10]

SECULAR EDUCATION OF THE WORKING CLASSES

The 1820s and 1830s were difficult times in Britain. Industrial and agricultural unrest led to riots, with mobs gathering at night to attack the homes of politicians and civic leaders. At times, revolution seemed close, especially as many soldiers sided with the general public and joined in demonstrations against the existing order. In June 1830, King George IV died and was succeeded by his brother, William IV, whose accession was followed by a general election in which parliamentary reform became the major issue. The Tory government of the Duke of Wellington was succeeded by the Whig administration of Earl Grey, with a reform platform. The existing electoral system, with its pocket and rotten boroughs[A] and its poor representation of Britain's industrial cities and towns, was no longer acceptable to the general public; neither was a right to vote that depended on inordinately restrictive financial criteria. Defeat of a Reform Bill at the committee stage in March 1831 led to another general election, which again returned a Whig majority. A second Reform Bill passed the House of Commons in September but was rejected by the House of Lords, leading to more riots, especially in Bristol and Nottingham. The (third) Reform Act was eventually passed in 1832 after initial

---

[A] Pocket boroughs were electoral districts controlled by one person or family by force or bribery, thus determining who was elected to parliament. Rotten boroughs were districts—often large in the past—that had declined to have only a tiny electorate but that still retained the right to parliamentary representation.

rejection by the House of Lords. It appeased the middle classes but left the working class without the vote, permitted the wealthy to continue to run the country, and accelerated the growth of trade unions. During this period, the church suffered a popular shift toward atheism and scientific enquiry, while the social order experienced a movement against privilege and aristocracy.

Henry Brougham, whose defense of Queen Caroline was referred to in Chapter 6, became the Lord Chancellor in 1830 in the administration of Lord Grey, and he had the satisfaction of seeing the passage not only of the Reform Act, of which he was a firm supporter, but also of an antislavery act. He had a major role in establishing various institutions for providing a secular education to the working, middle, and professional classes.

The Society for the Diffusion of Useful Knowledge was founded in 1826 by a group of educational reformers, and particularly by Brougham, to further the education of the general public and to make good books available cheaply to them. It started the Library of Useful Knowledge, a series of books published for as little as sixpence (one-fortieth of a pound) aimed at a mass readership, and included in this series was Bell's *Animal Mechanics* in two parts. Thirty thousand copies of these two sixpenny booklets sold within three years.[11] It was also hoped that Bell might publish anatomical plates cheaply for medical students through this vehicle, but the project was apparently abandoned.[12]

The Mechanics' Institutes, also founded in various British cities during and after the 1820s for educational and social purposes,[13] helped to teach technical and other subjects to working-class men through lectures, classes, books, and pamphlets and provided facilities for various socially acceptable activities (such as chess contests) away from the local bars and gin houses (Fig. 10.1). They helped to improve the efficiency of the workers and to reduce class differences and social unrest. It is notable that the first real Mechanics' Institute in Britain was probably the Edinburgh School of Arts,[B] founded in 1821 by Leonard Horner, later warden of the University of London (see Chapter 11). Many of these institutes evolved years later into regional universities or colleges, and some of their libraries later became public libraries. Interestingly, both the Society for the Diffusion of Useful Knowledge and the Mechanics' Institutes became increasingly concerned to maintain their objectivity and put at arm's length natural theology as a means to a scientific education. Indeed, the policy of the society and of its library became increasingly to avoid party politics and religion of all sorts so as to appeal to a wide audience and avoid controversy that might alienate its members. The society thus declined to publish the revision of Paley's work by Brougham and Bell; the Mechanics' Institutes recommended it for their libraries but stocked very few

---

[B] Now the Heriot-Watt University in Edinburgh.

WARWICK MECHANICS' INSTITUTION PERAMBULATING
LIBRARY.—SEE PRECEDING PAGE.

Figure 10.1 Mechanics' Institutes were founded in various British cities during and after the 1820s to educate the working classes and for social purposes. Illustration of the perambulating library of the Warwick Mechanics' Institution. (From *The Illustrated London News*, 1860; Courtesy of the Wellcome Library, London.)

other works on natural theology because of its potentially divisive nature and to avoid the appearance of bias or prejudice.[12]

THE BRIDGEWATER TREATISES

To many of the British middle and upper classes, Christian beliefs seemed to underlie an ordered society, the alternative being atheism and anarchy.[5] Such beliefs appeared to be challenged, however, by the scientific achievements that occurred in the first decades of the nineteenth century. This seeming decline in Christian beliefs may have been responsible for the Bridgewater legacy. Francis Henry Egerton, an eccentric clergyman and the eighth and last Earl of Bridgewater, died in 1829 and, in his will, directed that the sum of eight thousand pounds be used to publish one thousand copies of a work "on the power, wisdom, and goodness of God, as manifested in the creation." His had been an unusual life—his house in Paris, for example, was filled with cats and dogs, some of which he dressed fashionably as men and women, drove about in his carriage, and fed at his table.[14] He never married, had no offspring, and made a large bequest (of money and manuscripts) to the British Museum. The money given to support natural theology was a separate bequest to be held at the disposal of the president of the Royal Society (then David Gilbert, a politician and member of parliament), who requested the help of the Archbishop of Canterbury (William Howley) and the Bishop of London (Charles Blomfield) in proceeding further. The Bridgewater treatises were to summarize the contemporary state of science and to provide evidence for the existence of God based on the supposed accomplishments of the deity.

At first, a single volume was planned, with essays by several authors, but eventually this proved too restrictive. In the end, not one but eight treatises were commissioned to promote natural theology. The choice of authors proved difficult. Some put themselves forward or recommended their friends; however, the aim was not to reward individuals but, rather, to promote a concept. Among those considered was Herbert Mayo, Bell's former student and colleague, but Mayo was eventually turned down in favor of John Kidd (1775–1851),[c] who wrote on the physical constitution of humankind. A volume on the intellectual and moral dimensions of humans was authored by Thomas Chalmers (1780–1847), the Scottish theologian and reformer. John Herschel, the astronomer, was approached to write on astronomy, but he declined because of the "pecuniary reward." As he wrote to Gilbert, it was an opportunity for

> calling forth ... the talents of men, who with real science and irreproachable character, have their zeal chilled and their sphere of utility contracted by the "res angusta domi"[severe pressure of poverty]. To such persons ... who live, or rather starve on their science, but who prefer hunger in that good cause to competency in a less dignified calling, a thousand pounds (the sum you propose to devote to each of the eight sections of the work) would indeed be a more material and noble assistance.[5]

---

[c] Kidd was regius professor of physic in Oxford University and supposedly the first Oxford physician to abandon the wig, large-brimmed hat, and gold-headed cane as marks of his profession.

In the end, the astronomy volume was assigned to William Whewell (1794–1866), an Anglican priest and polymath.[D] There was general agreement among Gilbert and his two advisers that Charles Bell—whose views on natural theology were well known—would be an excellent choice to write on human anatomy, including the hand. Others selected to contribute were Peter Mark Roget (1779–1869), secretary of the Royal Society,[E] who wrote on physiology; William Buckland (1784–1856) on geology;[F] William Kirby (1759–1850)[G] on the habits and instincts of animals; and William Prout (1785–1850)[H] on chemistry and meteorology.

At times, Bell seemed bored by the project. As he wrote to his brother George, "And here are eight men more to wear the subject to the bone—all at the same work. You cannot wonder that I long to be at original matter and compositions—at science rather than writing."[15]

The series got off to a difficult start. Rumors of lobbying for authorship were countered by a published statement of facts. The selected authors received little advice from Gilbert about the intended scope of their contributions other than the general aim stated in the earl of Bridgewater's will and a list of the other contributors, and it seems that there was little communication between them.[16] The original publisher, John Murray, decided that the venture was too risky financially and backed out of it, while Longmans had concerns about the size of the intended market and offered unfavorable terms.[5,16] Contracts were finally signed with publisher William Pickering in April 1832, and the volumes appeared between 1833 and 1836, almost in serial form, because the authors failed to keep to the original schedule.[16] They were well written in a nontechnical but authoritative manner, were important contemporary scientific books,

---

[D] Whewell was a scientist, mathematician, moral philosopher, and theologian, and he is most remembered for his writings on the history and philosophy of science.

[E] One of the founders of the Medical and Chirurgical Society of London (now known as the Royal Society of Medicine), physiologist, and inventor of the log–log slide rule, Roget is best known for his thesaurus of English words and phrases, first printed in 1852. He taught for a while at the Great Windmill Street School on the theory and practice of physic.

[F] An English theologian, geologist, and paleontologist, Buckland was an eccentric who made a habit of eating exotic animals such as panther, mouse, and fly and is said to have eaten part of the mummified heart of King Louis XIV of France. He provided the first description of a dinosaur fossil, was professor of geology and mineralogy at the University of Oxford, and was dean of Westminster. His bust is in Westminster Abbey.

[G] Kirby was an English entomologist and country parson who coauthored a four-volume reference work on insects.

[H] Prout was a British physician and chemist who classified the content of food into fats, carbohydrates, and proteins; discovered the presence of hydrochloric acid in gastric juices; and developed a theory that the atomic weights of the elements were multiples of that of hydrogen. The proton is said to have been named after him, as also is the prout, a unit of nuclear binding energy.

and sold well, but they were written by eight individual authors of differing theological background (four were clerics), aimed at different audiences (the masses or the fashionable among the lay public; specialized or more general scientists and related professionals; and the devout, the theologians, or those seeking to reconcile their religious beliefs with a changing world), and did not really meet or respond to the various objections that could be made to the intelligent design argument.[5] The radical press pilloried the "Bilgewater" books.[17]

Bell's volume was well received; the initial print run was for one thousand copies, but more than thirteen hundred were ordered, and within eighteen months an additional seventy-five hundred copies had been printed. In his preface, Bell emphasized that "the reflexions contained in these pages have not been suggested by the occasion of the Bridgewater Treatises, but arose, long ago, in a course of study, directed to other objects."[18] In other words, he was expressing his own genuine beliefs rather than following a party line. Nevertheless, he also felt compelled to explain in the preface that "from at first maintaining that design and benevolence were every where visible in the natural world, circumstances have gradually drawn the author to support these opinions more ostentatiously and elaborately than was his original wish."[19] Bell probably chose to focus on the human hand because of its remarkable versatility and "the perfection" of its design. He traced the comparative anatomy of the hand back through monkey's paws, the paws of four-legged animals, the wings of birds, and the fins of fishes. He had no difficulty in accepting this continuity as a wonderful example of the way in which intelligent design led to the perfect adaptation of animals to their environment. In a sense, it can be said to have anticipated the "missing links" debates that occurred later in the nineteenth century.[20] After reviewing the actions of the muscles of the arm and hand, he turns to sensation, especially touch and the muscle (sixth) sense, and then the interaction of movement with sensation. The *Lancet* reviewed the book at length but in condescending and somewhat whimsical terms,[21] summarizing Bell's argument and quoting lengthy segments but failing to emphasize or credit some of the important new biological concepts that he spelled out for a lay public, such as the importance of sensation for skilled motor function (see Chapter 8).

It has been suggested that in his Bridgewater treatise, Bell connected in a single text his religious faith and anatomical background with his "concern for the machinery-mangled hands" of industrial workers,[20] but this seems somewhat far-fetched and implausible. Rather, it is more probable that he used the occasion to popularize and extend his scientific beliefs concerning movement and sensation in general by reference to the hand in particular, because of the hand's especial mechanical intricacy and importance to human life. In a way, he regarded the hand as a sense organ conveying touch, texture, contour—in the blind, the hand "sees"—and as an instrument in itself.

Bridgewater treatises conveyed "safe science"[12] and some—those by Whewell, Bell, Roget, and Buckland—were therefore recommended for inclusion in libraries of the Society for the Diffusion of Useful Knowledge.[12] The society continued until 1848, but with a shift away from natural theology. Manuscripts submitted to it for publication were reviewed by external referees before acceptance. Bell acted in this capacity with regard to a volume on physiology by Thomas Southwood Smith (1788–1861), a minister and physician, and objected to some of the materialist opinions expressed, leading ultimately to modifications of the offending passages.[12]

Southwood Smith, a pioneer in sanitary reform, believed that burying the dead was a waste of bodies that could be used for anatomical studies. In 1832, he achieved notoriety by publicly dissecting the body of his friend Jeremy Bentham, the social reformer and one of the founders of utilitarianism, who had died three days earlier at the age of eighty-four and left his body to medical science.[1] The Anatomy Act of 1832, which permitted the dissection of donated cadavers and those of unclaimed paupers by medical practitioners, anatomy teachers, and certain students, is said to have been passed as a direct result of an article by Smith. It helped to end the practice of grave robbing.

In 1837, a treatise written by Charles Babbage (1791–1871)—an English mathematician and developer of the difference engine, a prototype of the computer—was published in 1837 by John Murray, the man who had misguidedly turned down the opportunity to publish the eight Bridgewater treatises. This treatise is a work of natural theology that denies the existence of any conflict between science and religion. It was written in response to the eight original volumes and is not part of the series. It is nevertheless titled *The Ninth Bridgewater Treatise: A Fragment*. Babbage believed that creatures had emerged by successive special creations and emphasized the difficulty in interpreting the meaning, significance, or nuances of words and language, especially from the distant past, as provided in biblical accounts. He compared God to what would now be called a computer programmer, and the operation of the universe to a computer program, and he provided mathematical support for his views and against opposing beliefs. Just as a computer can be programmed to proceed according to one law for a certain number of operations and then to follow a different law, with the change being preprogrammed from the onset, so—he argued—new species

---

[1] Bentham's skeleton and a wax effigy of his head, dressed in his own clothes, can be seen in a glass-fronted case at University College London. His embalmed head is in a chest placed close by. An account of how this came about is provided by Marmoy CFA: The auto-icon of Jeremy Bentham at University College, London. *Med Hist* 1958; 2: 77–86. Legend (without factual basis) has it that Bentham is regularly wheeled into meetings of the College Council, where his presence is recorded in the minutes with the words "Jeremy Bentham—present but not voting."

could have been programmed by a deity at the onset (that is, at the creation) to appear after a set time. It was an interesting account, but it lost credibility a few years later as the theory of evolution by natural selection gained ground.

DECLINE OF NATURAL THEOLOGY WITH THE EVOLUTION OF EVOLUTION

This is not the place to discuss the popularity and subsequent decline in interest of natural theology or its modern equivalent, except in the briefest of terms. Proponents and critics abound, interpretations differ, and religious divisions still relate to the acceptance or otherwise of its tenets.[22] Natural theology had begun to decline in importance in the 1830s, and secularists scorned it as a measure of the "clerical hold over British science."[23] It had received support from the views of Cuvier in Paris, who held that successive creations had occurred after natural catastrophes. Cuvier's anatomical classification of animals was by *function*, and this led him to place animals into four distinct categories (as opposed simply into vertebrates and invertebrates), with no intermediate forms. He believed that animals could not evolve from one category to another because the constituent parts of an animal must operate together perfectly for maximal functional benefit. Thus, he held that piecemeal alterations in structure to allow evolution into new forms would impair the survival of the whole. However, Cuvier's views were being challenged, and exciting concepts of evolution were emerging.

Buffon (1707–1788) and other biologists and philosophers in the eighteenth century had begun to consider that life forms may not have been fixed at the time of the creation but, rather, evolved with time and geography. Erasmus Darwin (1731–1802)[J] wrote about the mutability of species, and in the early nineteenth century, Lamarck developed a theory that organisms evolve to survive in different environments (allowing acquired characteristics to be inherited) and that they become increasingly complex with time. Étienne Geoffroy Saint-Hilaire (1772–1844), also in France, noted that Cuvier's principles could not explain the structural similarities between different animal classes. Structural anatomy, he believed, more properly than function reflected the relationships between different groups of animals, with allowance made for compensatory changes such that overdevelopment of one anatomical feature might lead to compensatory alterations in others. He thus came to believe in the continuity of the animal kingdom. Richard Owen (1804–1892),[K] an English anatomist

---

[J] The philosopher–poet and physician was the grandfather of Charles Darwin.

[K] Owen—vain, jealous, and prickly—was a brilliant scientist who was instrumental in setting up what is now the Natural History Museum in South Kensington. He tutored Queen Victoria's children and eventually came to believe in evolution but opposed Darwin's belief that it was due to natural selection. It was he who recognized and named the dinosaurs as a distinct group.

and paleontologist, believed that all vertebrates had the same basic structure or "archetype" and showed that many structural similarities exist despite functional differences, as Geoffroy believed. These structural "homologies," which were held to reflect a common derivation, may not have had an immediate adaptive function, and reliance on function alone would therefore have neglected such morphological similarities. (By contrast, analogous structures have a similar function in different organisms but no common ancestral origin.) Owen's work on homologies provided support for the concept of evolution, an idea that was gaining ground with time, even with Owen himself.

One of the strongest supporters of the concept of evolution was Robert E. Grant (1793–1874) in London. Grant had obtained his MD degree from Edinburgh University in 1814, but he abandoned medicine for marine biology. He had been a member of the Plinian Society—a forum for discussing papers on natural history—at the university, which Charles Darwin—who became for a while Grant's student and assistant—also joined. After being appointed professor of comparative anatomy and zoology at the University of London (University College London) at approximately the same time as Bell, Grant—an evolutionist—remained there until his death, disseminating the views of Geoffroy (as opposed to Cuvier) and Lamarck. His atheistic views and nonconformist lifestyle made him enemies, among whom was Owen, who came to see him as a despised rival.[24] Grant must not have been very popular either with his colleague Bell, whose creationist views he was coming to overshadow.

The subsequent publication anonymously of the controversial *Vestiges of the Natural History of Creation*[L]—a best-seller full of facts, speculations, and unsupported generalizations that popularized science but eventually aroused the enmity of both the scientific community and the church—was a further blow to natural theology and gave greater and wider appeal to the concept of evolution. The general thesis of the book was that species are not immutable. Many of the lines of supportive evidence used in *Vestiges* were the same as Charles Darwin had been accumulating,[25] influenced as he was by the geologist Charles Lyell,[M] who is said to have taught him "how to think about nature."[26]

[L] Initially published anonymously in 1844, its author was later identified as Robert Chambers (1802–1871), a Scottish journalist and publisher. A detailed account of the book and its impact can be found in Secord JA: *Victorian Sensation: The Extraordinary Publication, Reception, and Secret Authorship of "Vestiges of the Natural History of Creation."* University of Chicago Press: Chicago, 2000.

[M] Charles Lyell (1797–1875), author of the famous multivolume *Principles of Geology*, believed that gradual environmental processes—rather than catastrophic events—led to major changes in the animate and inanimate world. He initially had difficulty in accepting the concept of evolution despite his friendship with Charles Darwin, and he was even more hesitant in accepting the mechanism of natural selection.

The effect of *Vestiges* on Darwin was dramatic, and—as criticism (as well as praise) for the book accumulated—he began doubting his own work and judgment.[25] It was to be several years before Darwin—suddenly aware that others had arrived at similar conclusions to his own—published the idea of natural selection as an evolutionary mechanism, jointly with Alfred Russel Wallace, in the *Proceedings of the Linnean Society* in 1858.[27] Their joint publication was engineered by Lyell and the botanist Joseph Dalton Hooker (1817–1911). Darwin's book, *On the Origin of Species*, was published in the following year, and in it Darwin attacked *Vestiges* as unreliable, hoping to distance himself and his work from the criticism it had generated.

Ironically, Darwin had once been a creationist and a follower of natural theology, but his views had changed with time, especially as he came to see how the remarkable adaptation of form to function that existed in nature could be explained by natural selection and did not require belief in an intelligent designer. Thus, the natural theology that Bell and Brougham espoused began to seem increasingly tired and dated in an era of change and reappraisal, failing to provide answers that were needed to satisfy an increasingly discriminating and questioning society. The decline in popularity of natural theology among students and intellectuals almost certainly had an adverse impact on Bell's standing as a teacher and university professor. He continued to frame his thoughts and anatomical lectures on the concept of intelligent design, thereby limiting their appeal and appearing increasingly out of touch with advances in the field.

NATIONAL HONORS

Bell's busy and successful life continued in London in the 1820s and early 1830s, where he became one of the founders and first professors of the new University of London, until he became disillusioned and dissatisfied with the way in which the new institution was managed, as discussed in Chapter 11: "The University, which was wont to be a subject of our correspondence, is going fast to the dogs; misrule and mismanagement are doing their work most efficiently."[28] He seemed indifferent to the Reform Bill, "which will do no earthly good,"[29] but was unable to leave the troubled metropolis because of his dependence on the income derived from his increasing clinical practice. He was losing his drive and interests: "Little objects of ambition which so long kept me afloat and struggling in this great tide, have no longer any attraction for me."[29] His work on natural theology and on *The Hand* kept him busy. He mixed with the privileged in society and entertained the intellectual leaders of the period, including Cuvier, who dined with him.[30] He relaxed when he could by visits to Scotland and the countryside and by fishing, which he enjoyed increasingly.

Bell had been a fellow of the Royal Society of Edinburgh since 1808.[N] In 1826, he was elected a fellow of the Royal Society in London,[O] and three years later he received its Gold (Royal) Medal for his discoveries relating to the nervous system (Figs. 10.2 and 10.3). At the Royal College of Surgeons, he served on the Council from 1830 to 1836. In 1832, he was elected Hunterian professor of comparative anatomy at the college (having served previously as Hunterian professor of surgery for the four years 1825 to 1828; see p. 91) and delivered a series of fifteen lectures on the Hunterian preparations that was published in the *Lancet*.[31] The attendance at least initially

> was scanty as might have been expected from the fallen character of this institution. . . . Sir Charles delivered it [his address] in a very admirable manner, receiving the warm plaudits of his auditors at the close. . . . On the table were skeletons of the porpoise, the antelope, the armadillo, the monkey, the manis . . . &c."[32]

The University of Göttingen conferred the degree of MD on Bell and his friend Astley Cooper in 1833, and he was elected a foreign member of the Royal Swedish Academy of Sciences in 1841. But the honor most prized was a knighthood.

The Royal Guelphic Order of Knighthood[P] was established in 1815 by the Prince Regent (later King George IV) on behalf of his ailing father, King George III. It was a Hanoverian rather than a British order, administered by the Hanoverian state, which became a kingdom at the Congress of Vienna after the Napoleonic Wars and was ruled by the British sovereign until 1837.[Q] The order was intended to reward meritorious service by officers of the Hanoverian regiments that were part of the British Army during the Napoleonic campaigns and also to reward British subjects who had performed important services to Hanover or to the Prince Regent himself. There were three levels of award—knight grand cross, knight commander, and knight—with the two top levels being reserved for senior military and government personnel. Attendance at a

---

[N] Bell was proposed for election in June 1807 and elected in June 1808. But he seems also to have been proposed in December 1810, raising the possibility that he was elected in 1811, not 1808. The records are unclear, but they are summarized by Waterston CD, Shearer AM: *Biographical Index of Former Fellows of the Royal Society of Edinburgh 1783–2002, Part 1, A–J*. Royal Society of Edinburgh: Edinburgh, 2006.

[O] His proposers were William Maton, William Blizard, Charles Konig, Everard Home, John Knowles, Henry Kater, Charles Babbage, James McGrigor, Charles M. Clarke, Henry Cline, Joseph Henry Green, John Bostock, Robert Brown, Grant David Yeats, Thomas Murdoch, Gilbert Blane, John Ayrton Paris, John Yelloly, Thomas Grey, and John Cooke.

[P] The Guelphic Order took its name from Guelph, the old dynastic name of the house of Hanover.

[Q] Queen Victoria, who succeeded to the British throne in 1837, could not become sovereign of Hanover because of her gender.

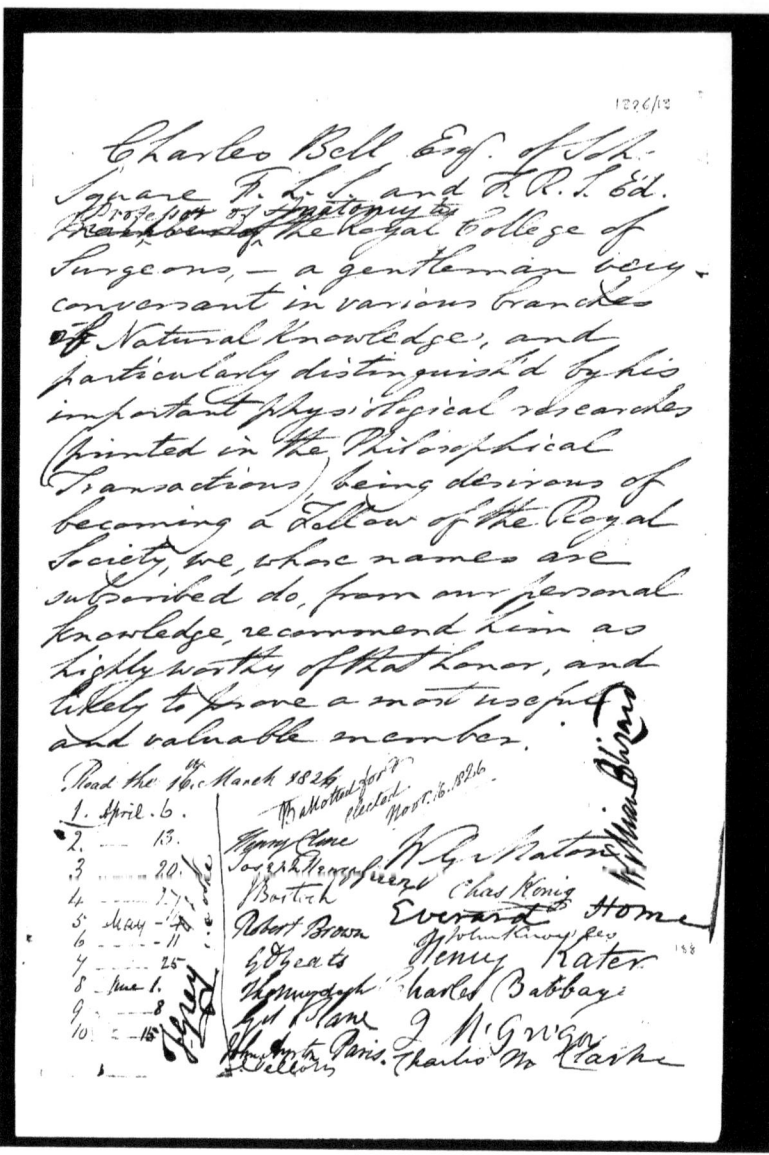

Figure 10.2 Certificate of election of Charles Bell to the Royal Society. © The Royal Society. His twenty nominators included seven physicians, six surgeons, an obstetrician, and a merchant, botanist, naturalist, writer, physicist, and mathematician. He was elected on 16 November 1826.

levee allowed the Prince Regent to invest with the insignia those appointed to the order. Britons appointed to the order were not normally given the title of "Sir" unless actually knighted by the sovereign.[33]

Lord Brougham—Bell's old friend—had become lord chancellor in 1830 and was anxious to promote the status of science in Britain by honoring its leading figures, a policy with which both King William IV and the prime

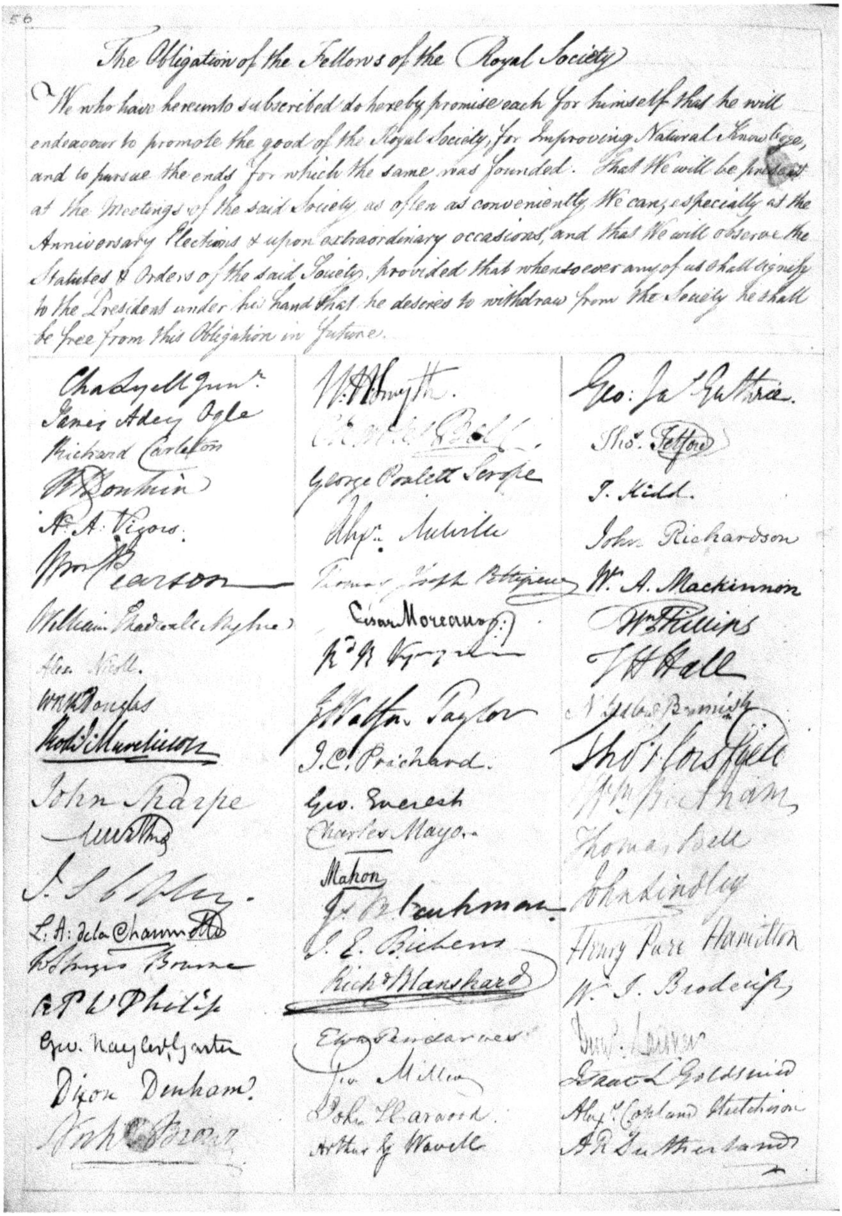

The Obligation of the Fellows of the Royal Society

We who have hereunto subscribed do hereby promise each for himself that he will endeavour to promote the good of the Royal Society, for Improving Natural Knowledge, and to pursue the ends for which the same was founded. That We will be present at the Meetings of the said Society as often as conveniently We can, especially at the Anniversary Elections & upon extraordinary occasions, and that We will observe the Statutes & Orders of the said Society, provided that whensoever any of us shall signify to the President under his hand that he desires to withdraw from the Society he shall be free from this Obligation in future.

Chas Lyell Junr.
James Adey Ogle
Richard Carleton
N. A. Vigors.
Wm Pearson
John Sharpe
Dixon Denham

W. H. Smyth.
Charles Bell
George Poulett Scrope
J. C. Prichard.
Geo. Everest
Charles Mayo.
Mahon
Rich. Blanshard
Arthur G. Wavell

Geo: Ja: Guthrie.
Thos. Telford
J. Kidd.
John Richardson
W. A. Mackinnon
Wm Phillips
Thomas Bell
John Lindley
Henry Parr Hamilton
W. J. Broderip
Isaac L. Goldsmid

Figure 10.3 Charter book of the Royal Society, page 56, with Charles Bell's signature to the obligation of fellows, middle row second from the top. © The Royal Society. It is beneath the signature of William Smyth, admiral and hydrographer, and in the same column as that of the surveyor-general of India whose name was given to the highest mountain in the world (George Everest), and—among others—those of an expert in ancient Egyptian mummies, an Anglo-Saxon scholar, and an English soldier of fortune. On the same page are the names of the geologist Charles Lyell, who so influenced Charles Darwin; the financier and philanthropist Isaac Goldsmid; and the civil engineer and bridge-builder, Thomas Telford.

Figure 10.4 Portrait of Sir Charles Bell in court dress and wearing the insignia of a Knight of Hanover. Engraving from *Medical Portrait Gallery* by Thomas Joseph Pettigrew, London, 1838–1840. (Courtesy of the Wellcome Library, London.)

minister, Lord Grey, agreed. At the time, official honors were usually awarded only to those in government or military service. It was finally decided to use the third class of the Guelphic Order to reward intellectual and scientific achievement, and in October 1831, it was announced that seven eminent British scientists were to be so honored. These were Charles Bell; Charles Babbage, the mathematician who authored the so-called Ninth Bridgewater Treatise (who declined the award); John Herschel, the astronomer; John Ivory, the mathematician; David Brewster, mathematician, physicist, and astronomer; John Leslie, another mathematician and physicist; and Charles Konig, mineralogist and botanist. It can be no accident that four of these were Scottish. Charles Bell was thus admitted to the Guelphic Order.

He was knighted by King William on 12 October 1831 at St. James's Palace in London (Fig. 10.4). Only he and Herschel of the six were there.[R] To his wife, he looked quite magnificent in his borrowed Court dress—"dark-brown and cut steel, white satin waistcoat, chapeau, sword, buckles, point d'Alencon frills and cuffs, and his Order with the Blue Ribbon."[34] The new knight described the investiture:

> I persuaded Herschel that on this occasion he represented the higher sciences, and that, therefore he must precede me in receiving the accolade, and he did precede me into the presence chamber, but in approaching the lord in waiting he lost heart, and suddenly countermarched, so that I found myself in front. My niece's dancing master having acted the king the night before, I had no difficulty.[35]

The *Lancet* commented, "There is not in England an anatomist or physiologist toward whom the favour of the sovereign could have been bestowed with more propriety."[36]

His servants now began addressing him as Sir Charles, somewhat to his embarrassment, but he dared to hope that his new title would teach them greater civility. In 1832, he moved from Soho Square to 30 (now, 82) Brook Street, close to Grosvenor Square, as if to live up to his new stature and title.

REFERENCES

1. Bell C: *An Essay on the Forces Which Circulate the Blood; Being an Examination of the Difference of the Motions of Fluids in Living and Dead Vessels.* p. vii. Longman/Burgess and Hill: London, 1819.
2. Anon: Introductory anatomical lectures: Mr. Charles Bell. *Lancet* 1825; 5: 99–102.
3. Desmond A: *The Politics of Evolution: Morphology, Medicine, and Reform in Radical London.* pp. 111–112. University of Chicago Press: Chicago, 1989.
4. Anon: Editorial [on Mr. Green on the King's College, and the three special providences.] *Lancet* 1832; 19: 151–155.
5. Brock WH: The selection of the authors of the Bridgewater treatises. *Notes Rec R Soc Lond* 1966; 21: 162–179.
6. Gardner-Thorpe C: William Paley (1743–1805): Neuroanatomist? *J Med Biogr* 2002; 10: 215–223.
7. Bell C: *The Hand: Its Mechanism and Vital Endowments as Evincing Design.* Preface, p. x. Pickering: London, 1833.

[R] The knighthood posed particular problems for John Herschel because his father, the famed William Herschel, had been appointed by the Prince Regent as a knight of the Royal Guelphic Order in 1816 but had never been formally knighted. He used the prefix "Sir" but properly was not entitled to do so. The implications of this for the family are discussed in Hanham A, Hoskin M: The Herschel knighthoods: Facts and fiction. *J Hist Astron* 2013; 44: 149–164.

8. Paley W, Brougham H, Bell C: *Paley's Natural Theology, with Illustrative Notes, to Which Are Added Supplementary Dissertations, by Sir Charles Bell.* 2 vols. Knight: London; Jackson: New York, 1836.
9. Desmond A: *The Politics of Evolution: Morphology, Medicine, and Reform in Radical London.* p. 203. University of Chicago Press: Chicago, 1989.
10. Bell C: *Animal Mechanics: Or, Proofs of Design in the Animal Frame: The Perfection of Design in the Bones of the Head, Spine, and Chest, Shown by Comparison with Architectural and Mechanical Contrivances.* Sine nomine: London, 1827; Reprinted in *Animal Mechanics*, by Charles Bell, Jeffries Wyman, Morill Wyman. pp. 90–92. Riverside Press: Cambridge, MA, 1902.
11. Anon: Art. VII. 1. The Nervous System of the Human Body. By Sir Charles Bell, K.G.H., F.R.S. 3rd edition. 1836; 2. The Hand; Its Mechanism and Vital Endowments, as Evincing Design. By Sir Charles Bell. 4th edition, 1837; 3. Narrative of the Discoveries of Sir Charles Bell in the Nervous System. By Alexander Shaw, Surgeon to the Middlesex Hospital. London, 1839. *Lond Quart Rev* (American edition) 1843; 72: 103–124.
12. Topham J: Science and popular education in the 1830s: The role of the Bridgewater Treatises. *Br J Hist Sci* 1992; 25: 397–430.
13. Kelly T: The origin of Mechanics' Institutes. *Br J Educ Studies* 1952; 1: 17–27.
14. Sutton CW: Egerton, Francis Henry. Dictionary of National Biography, 1885–1900, Vol. 17. Retrieved from https://en.wikisource.org/wiki/Egerton,_Francis_Henry_(DNB00); accessed 20 September 2015.
15. Bell C: Letter to George dated 3 September 1881 (*sic*). p. 320. In *Letters of Sir Charles Bell, K.H., F.R.S. L. & E: Selected from His Correspondence with His Brother George Joseph Bell.* Murray: London, 1870.
16. Topham JR: Beyond the "common context": The production and reading of the Bridgewater Treatises. *Isis* 1998; 89: 233–262.
17. Desmond A: *The Politics of Evolution: Morphology, Medicine, and Reform in Radical London.* p. 20. University of Chicago Press: Chicago, 1989.
18. Bell C: *The Hand: Its Mechanism and Vital Endowments as Evincing Design.* p. IX. Pickering: London, 1833.
19. Bell C: *The Hand: Its Mechanism and Vital Endowments as Evincing Design.* p. XI. Pickering: London, 1833.
20. Capuano P: On Sir Charles Bell's *The Hand*, 1833. In Felluga DF (ed): *BRANCH: Britain, Representation and Nineteenth-Century History. Extension of Romanticism and Victorianism on the Net.* Retrieved from http://www.branchcollective.org/?ps_articles=peter-capuano-on-sir-charles-bells-the-hand-1833; accessed 13 September 2015.
21. Anon: The hand: Its Mechanism and Vital Endowments as Evincing Design. By Sir Charles Bell, K.G.H., F.R.S. L. & E. Aldi Discip. Anglus. London. Pickering.1833 8vo. *Lancet* 1833; 21: 165–169.
22. Chignell A, Pereboom D: Natural theology and natural religion. In Zalta EN (ed): *Stanford Encyclopedia of Philosophy.* Fall 2015 edition. Retrieved from http://plato.stanford.edu/archives/fall2015/entries/natural-theology; accessed 27 September 2015.
23. Desmond A: *The Politics of Evolution: Morphology, Medicine, and Reform in Radical London.* p. 379. University of Chicago Press: Chicago, 1989.

24. Desmond A: *Archetypes and Ancestors: Palaeontology in Victorian London, 1850–1875.* pp. 116–119. University of Chicago Press: Chicago, 1984.
25. Browne J: *Charles Darwin: Voyaging. Volume I of a Biography.* pp. 457–472. Cape: London, 1995.
26. Browne J: *Charles Darwin: Voyaging. Volume I of a Biography.* pp. 186–190. Cape: London, 1995.
27. Darwin CR, Wallace AR: On the tendency of species to form varieties; And on the perpetuation of varieties and species by natural means of selection. *J Proc Linnean Soc Lond Zool* 1858; 3: 45–62.
28. Bell C: Letter to George dated 18 February 1831. pp. 316–317. In *Letters of Sir Charles Bell, K.H., F.R.S. L. & E: Selected from His Correspondence with His Brother George Joseph Bell.* Murray: London, 1870.
29. Bell C: Letter to George, undated 1831. pp. 318–319. In *Letters of Sir Charles Bell, K.H., F.R.S. L. & E: Selected from His Correspondence with His Brother George Joseph Bell.* Murray: London, 1870.
30. Bell C: Letter to George dated 12 August 1830. pp. 313–314. In *Letters of Sir Charles Bell, K.H., F.R.S. L. & E: Selected from His Correspondence with His Brother George Joseph Bell.* Murray: London, 1870.
31. Bell C: Lectures on the Hunterian preparations in the museum of the Royal College of Surgeons, London. *Lancet* 1833; 21: 279–285, 313–319, 486–492, 912–919, 962–969; and *Lancet* 1834; 22: 216–221, 265–271, 346–352, 410–416, 745–751, 794–806, 824–829, 875–887 (multiple parts).
32. Anon: Lectures at the London College of Surgeons. *Lancet* 1833; 19: 768.
33. Hanham A, Hoskin M: The Herschel knighthoods: Facts and fiction. *J Hist Astron* 2013; 44: 149–164.
34. Bell M: Letter to Miss Bell dated 12 October 1831. p. 323. In *Letters of Sir Charles Bell, K.H., F.R.S. L. & E: Selected from His Correspondence with His Brother George Joseph Bell.* Murray: London, 1870.
35. Bell C: Notes. 12 October 1831. pp. 324–325. In *Letters of Sir Charles Bell, K.H., F.R.S. L. & E: Selected from His Correspondence with His Brother George Joseph Bell.* Murray: London, 1870.
36. Anon: Sir Charles Bell. *Lancet* 1833; 20: 756–761.

11

NEW CLASSROOMS: OLD STRUGGLES

Charles Bell was a renowned teacher of anatomy and surgery and an eloquent advocate of natural theology (see Chapter 10). In addition to his many surgical and anatomical textbooks and his original researches, which earned him an international reputation as clinician and scientist, his *Essays on the Anatomy of Expression in Painting* had a major influence on art and on the training of painters and artists, and it was instrumental in the birth of physiological psychology (see Chapter 4). Bell founded his own anatomy school, took over and directed the well-known Hunterian school in Great Windmill Street, was much sought after as a clinical teacher at one of the great London hospitals, was the most famous professor appointed to the medical department when the University of London was established, helped to found the Middlesex Hospital Medical School in the capital, and held a prestigious chair of surgery at the University of Edinburgh. And yet, through no fault of his own, he did not follow through at either the University of London or the Middlesex Hospital Medical School, leaving after such a short time that he could have put the existence of these institutions in jeopardy or, at the very least, harmed their chances of success. Bell's involvement in British medical education and its reform is considered in this chapter.

THE BACKGROUND

The royal hospitals in London during the eighteenth century consisted of St. Thomas's and St. Bartholomew's, which cared for the sick; the Bethlem, for the insane; Christ's Hospital, for foundling children; and Bridewell, which provided shelter and training for homeless children. Both St. Thomas's and St. Bartholomew's were general hospitals that had been closed with the dissolution of the monasteries in the 1530s but re-established soon after by royal charter. They were wealthy with endowments, as was Guy's Hospital, founded in 1721 and built opposite St. Thomas's at the express wish of its founder, a wealthy publisher and speculator. Other voluntary hospitals were founded in London during the following years and evolved into institutions for curing as well as caring for the sick, and these included the Middlesex Hospital, founded to treat the sick and lame of Soho (see p. 84).

Nurses were appointed by an all-powerful matron, but they often came from poor families and functioned simply to watch over the sick, keeping them clean, dry, and provided with food and medicine, but with limited compassion or understanding of their disorders. The ideal nurse was strong, sober, clean, and— it was hoped—literate, but the ideal was seldom encountered, perhaps because nurses were often paid no more than domestic servants. Wards were under the charge of a ward sister, who was responsible to the matron.[1] During the nineteenth century, the established hospitals of the metropolis gradually assumed an additional function as teaching and research centers, with lectures given by the physicians and surgeons appointed there, and in this way they gradually took over the function of the private schools referred to in Chapter 3.[2]

At that time, only a small proportion of medical men were university-trained physicians, the others being surgeons (who trained by apprenticeship and then gained membership of the College of Surgeons), apothecaries (who worked in general practice), or surgeon–apothecaries. It has been suggested that the unregulated and decentralized system of private medical education in London was successful, flexible, and comparable to that available in Edinburgh or Paris,[3] but this seems at odds with its unplanned structure, the limited experience and training obtained by many students, the lack of widely accepted standards, and the nepotism that allowed individuals of limited abilities to flourish as teachers.

Various attempts to reform the medical profession in the early nineteenth century had met with resistance from competing interests until—in 1815—the Apothecaries' Act was passed by parliament (despite opposition from physicians and surgeons[4]) and came to regulate aspects of medical practice for the next forty years. It required those without a university degree to sit for the license of the Apothecaries' Hall if they intended to work as a general practitioner or apothecary. A curriculum was established; standards were set; and both hospitals and dispensaries (outpatient centers, generally supported by charity or subscription) were utilized for clinical teaching, with attendance required for at least six months, later extended to nine months.[5] A five-year apprenticeship to an apothecary and testimonial letters were needed, as well as evidence of competence in Latin and attendance at two courses on anatomy and physiology and two on the theory and practice of medicine.[5-7]

The College of Surgeons had stricter requirements of candidates for its diploma. In 1824, for example, candidates were required to have spent at least six years in acquiring professional knowledge, involving regular attendance of at least three winter courses of anatomical studies and one or more winter courses of surgical studies, with performance of dissections during two or more of these winter courses, and attendance at the surgical practice at a hospital for at least one year.[7,8]

To meet the requirements of the Society of Apothecaries and the College of Surgeons, prospective candidates typically spent time in London, selecting courses of instruction by teachers at various institutions. It was common, for example, to attend anatomy classes at a private school and then to gain experience of clinical surgery in a hospital and of clinical medicine in a dispensary. These dispensaries played an important role in medical education, especially after 1815,[9] despite being vilified by Thomas Wakley in the pages of the *Lancet*.[5] The court of examiners of the Society of Apothecaries went on to develop careful regulations for the recognition of dispensaries as educational establishments, including the requirement that they be affiliated with a recognized medical school.[5] For the benefit of trainees, a general listing of the medical, surgical, and anatomical schools in the capital was published in 1825, together with brief commentary on the quality of education provided.[10]

During the 1820s and 1830s, as hospitals became more important as teaching institutions, the influence of the private schools diminished. Their demise was hastened by the creation of the University of London, which offered well-organized detailed courses and weekly examinations, and which hoped to make attendance at systematic lectures the prime part of the early years of training, with clinical work postponed to the later years. In this way, a predominantly clinical apprenticeship was replaced by a more solidly based, scientific training.

THE UNIVERSITY OF LONDON

Until 1826, the universities of Oxford and Cambridge were the only ones in England, and entry to the former or graduation with a degree from the latter was subject to certain religious tests. The suggestion to set up a secular university in London followed an open letter from the poet, Thomas Campbell,[A] addressed to Henry Brougham and published in *The Times* on 9 February 1825. Isaac Goldsmid (1778–1859), a financier who actively supported the founding of the new university, had introduced Campbell to Brougham.[B] A committee was set up to explore the possibility further, and the University of London was established by deed of settlement in February 1826.

[A] Thomas Campbell (1777–1844) was a Scottish poet whose "Pleasures of Hope" brought him fame. He was involved in the foundation of the University of London and was elected Lord Rector of Glasgow University (1826–1829) in competition against Sir Walter Scott. He is buried in Westminster Abbey at Poet's Corner.

[B] Goldsmid also helped to establish University College Hospital in 1834, eventually becoming its treasurer, and he was active in educational and penal reform. In addition, he worked hard for the emancipation of Jews in the United Kingdom, with success following final passage of a bill through the House of Lords in 1858, due in part to his efforts. In 1841, he became the first Jewish baronet.

Capital was raised by selling shares, land was purchased, and buildings were erected (in and around Gower Street) despite the hostility of the Anglican Church and the Tories. The foundation stone was laid at the end of April 1827 using the same mallet[C] as had been used by King Charles II for the first stone of St. Paul's Cathedral.[11] This was followed by a celebratory dinner attended by Bell at the Freemasons' Tavern in Great Queen Street (Lincoln's Inn Fields), which raised eight thousand pounds for the project.[12] From the shareholders (or proprietors,[D] as they were called), twenty-four persons were elected to a governing council, and they appointed Leonard Horner as warden in May 1827.[E] He—like Bell and Brougham—was from Edinburgh, and the three of them were on good terms. Thus it happened that Bell was connected unofficially but from a very early stage with the developing university (Fig. 11.1). The first professors, including Bell, were selected by personal invitation to occupy a chair; later appointments were advertised and required formal application.[13] Professorial appointments first began to be confirmed in the summer, but confusion followed with regard to the position offered to Bell.

Bell had believed that "he was to be considered the head of the medical school" but then heard that the council was attempting to procure a European professor and, finally, that the council had elected Granville Sharp Pattison (1791–1851) as professor of anatomy. Because this chair in other schools was occupied by the most senior person, Bell immediately announced himself released "from all connection with the institution."[14] He was reassured that this was all a misunderstanding and that he was to take the "higher department of anatomy."[11] Indeed, with his acquiescence, it was initially announced that Bell, Johann Friedrich Meckel,[F] and Pattison would be appointed jointly to the chairs

---

[C] The mallet had been presented by Sir Christopher Wren, the cathedral's architect, to the Masonic Lodge of Antiquity to which he belonged.

[D] The privileges of the proprietors included the right to nominate one student for each share (of one hundred pounds) held.

[E] Leonard Horner (1785–1864), the younger brother of Francis Horner (see Chapter 5), was a Scottish geologist, an educational reformer, founder in 1821 of the School of Arts in Edinburgh to provide technical education for the working classes, and one of the founders of the Edinburgh Academy. In 1828, he became warden (supervisor) of London University, resigning in 1831. In 1833, after spending time in Bonn to rest, he was appointed to the Royal Commission on the Employment of Children in Factories, and he served for many years as Inspector of Factories, doing much for the welfare of women and children. A more detailed account is provided by Brown CM: Leonard Horner, 1785–1864: His contribution to education. *J Educ Admin Hist* 1985; 17: 1–10.

[F] Johann Friedrich Meckel (1781–1833) was an anatomist, pathologist, and zoologist at the University of Halle, Germany. He worked with Geoffroy, Cuvier, and von Humboldt, becoming an evolutionist and ultimately a follower of Lamarck. He believed that fetal deformities relate to premature arrest of development, relating in turn to lower forms of life.

Figure 11.1 Bell was the most famous of the professors appointed to the medical department when the University of London was established. The illustration shows the main building of the original university (now University College London) in Gower Street. Engraving by W. Wallis after T. H. Shepherd, 1828. (Courtesy of the Wellcome Library, London.)

of anatomy, surgery, and physiology, but then Meckel withdrew. Bell wrote to George on 2 January 1828:

> We make no very decided progress in the University. I am thinking of taking the physiology and the clinical surgery. I have my choice. . . . They are the courses which I know will be most generally interesting—the one elegant—touching on whatever I choose—the other interesting to all ages.[15]

When Bell was offered just the chair of physiology, he protested that he had not lost interest in "practical subjects" and requested appointment also to the chair of surgery or clinical surgery.[16] He preferred the clinical appointment so that he could serve both his hospital and the university without the duties becoming too onerous. It was therefore agreed that the teaching of physiology and clinical surgery would be left to him and that of anatomy and morbid anatomy to Pattison.[17] And this is indeed what Bell accepted,[G] allowing the university to benefit from his already established reputation. The appointment of Pattison, however, was to lead to enormous trouble.

Bell had to ensure that his course on surgery was sufficiently comprehensive that it satisfied the requirements of the College of Surgeons, an action that later was construed to indicate that he had also been appointed to the vacant chair of surgery.[17] According to Bell, however, he was told that it had been "the universal wish of my colleagues that I should be professor of surgery," a situation that he supposedly did not welcome because it increased his workload, requiring him to lecture every day.[14] Regardless, in published notices announcing forthcoming lectures and courses, Professor Bell is listed as lecturing on both surgery and clinical surgery.

A statement subsequently published by the council provided a syllabus to the courses offered by the various professors, including those by Robert E. Grant (discussed in Chapter 10) on comparative anatomy and zoology, Pattison on anatomy and morbid anatomy, and Bell on physiology and clinical surgery.[18] Bell later stated before a parliamentary commission that he believed that "the teaching of physiology, as well as the practice of surgery, cannot be separated from anatomy, without becoming diffuse, verbose, and of little use."[19] Accordingly, he considered himself to be appointed to lecture on subdisciplines of anatomy. Although it has sometimes been suggested that—when the institution actually opened—he was also given the chair of anatomy and morbid anatomy,[20] there is no evidence of this. To the contrary, he wrote to the council in August 1831 requesting that it state to the proprietors that he had "never been a candidate for the professorship of anatomy"[21] to quell further gossip on the topic.

[G] This cost Bell four hundred pounds (see Chapter 3, footnote R).

The new university first opened its doors to students in October 1828. The medical department was a very active part of the university from the beginning, with one hundred and sixty-five students enrolling in the first year.[22] Indeed, in its early years, the university is said to have owed its very survival to the success of its medical school.[23] There were classes in chemistry, botany, comparative anatomy, anatomy and operative surgery, physiology, the nature and treatment of disease, clinical medicine, surgery and clinical surgery, midwifery and diseases of women and children, and materia medica and pharmacy.[18] Prices ranged between three and five pounds for each course.[24]

With Bell on staff and his colleague from the Middlesex Hospital, Thomas Watson, appointed to the chair of clinical medicine,[H] a special clinical link existed with that hospital, and it was hoped by several of the founders that the Middlesex would serve as the clinical arm of the university. The premature disclosure of this aspiration as a fait accompli in an advertisement by the council of the university, published in *The Times* on 3 September 1828, without their knowledge, infuriated the governors of the hospital, who immediately distanced themselves from the notice and from any formal association with the university. Concern about reduced financial support of the hospital because of its association with the secular university may also have influenced them, and subsequent attempts by the university to link with the hospital were rebuffed.

The fee structure at the time shows the special relationship existing between the new university and the Middlesex Hospital, which offered clinical instruction but had no formal school of its own. The usual fee structure for teaching at the Middlesex was as follows: for a physician's pupil, ten guineas[I] for six months, fifteen guineas for twelve months, and twenty-one guineas for a lifetime; for a surgeon's pupil, the corresponding charges were fifteen, twenty, and fifty guineas. By contrast, for hospital (medical or surgical) practice during an academic session of nine months, students of the university were charged twelve guineas; a second session also cost twelve guineas, with free admission thereafter.[22] However, student activists claimed anonymously—perhaps because they feared retribution—that the cost of a medical education was significantly greater at the university than elsewhere in London, and it was left to anonymous professors to counter their claims.[25]

[H] Sir Thomas Watson (1792–1882) was elected physician to the Middlesex Hospital in 1827 (resigned in 1843) and appointed professor of clinical medicine at the University of London for a year before transferring to King's College as professor of forensic medicine and then of the principles and practice of medicine (resigned in 1840). He was later appointed physician extraordinary to Queen Victoria. He described the water hammer pulse of aortic regurgitation ("Watson's sign").

[I] A guinea was one pound and one shilling. There were twenty shillings in a pound. At that time, the pound was valued at between four and five American dollars.

Charles Bell, perhaps the most famous of the medical faculty, delivered the opening inaugural lecture at the university on 1 October 1828. He had prepared it while fishing in Wales "and dictated it in Aberystwith [sic] to the applauding hands of my wife":[26]

> The success of the institution will depend on the relations established between the teacher and the pupil; the devotion of the one to the interests of science and to the instruction of youth, and the gratitude of the other for the highest benefit that can be received—the improvement of his mind and the acquisition of a profession. . . .
>
> If I value highly the influence of this great establishment, it is because I have been long engaged in teaching, and have experienced all the difficulties of forming a medical school. . . .
>
> In colleges, such as have been instituted in former ages, the professors enjoy the advantages of independence and seclusion, and are removed from the distractions of our busy world: it is otherwise in London. Here, professional men are differently situated, and more activity is requisite—perhaps of a different kind; less contemplative or theoretical—more practical; and to maintain a distinguished place increasing exertion is necessary. . . .
>
> In the course of thirty years I have seen the establishment of many schools attempted, but it has always happened that the temptation of following a lucrative practice has far outweighed the desire of reputation to be gained by learning; and, consequently, just when the professor becomes useful by the knowledge he was capable of communicating, he has withdrawn himself; and so the situation of a medical teacher, instead of being the highest . . . is merely looked upon as a situation introductory to business. . . . Let us hope that, instead of this rapid succession, this university may be able to raise the professors of science to higher considerations. . . .
>
> With respect to our students, the defects of their mode of education are acknowledged on all hands. They are at once engaged in medical studies without adequate preparation of the mind; . . . without having acquired the habit of attention to a course of reasoning; nor are they acquainted with those sciences which are really necessary to prepare them for comprehending the elements of their own profession. But in this place this is probably the last time they will be unprepared. . . .
>
> [T]here is something peculiar in the character of the medical student. His occupations lead him off from authority at an immature and dangerous time of life: the pursuit of experimental philosophy, and the history of those sciences to which he is now introduced, tend to give him a mean opinion of the efforts of individuals, and to beget a suspicion of any thing like authoritative language. . . . To our students books are no longer talismans and spells; they have no respect for antiquity, and names have no authority for them. Taught to surrender their judgment to experiment only, can we be surprised that they require to be reasoned with?[27]

After a few other reflections, including comments on religion and worship that must have seemed out of place in the secular environment of Gower Street, Bell concluded his introductory lecture and went on to lecture on the circulation, pointing to its dependence on the principles of hydraulics as evidence of design in the animal frame. Clearly he had no intention of changing the focus of his lectures based on the politics of the new institution. Other members of the medical department (later renamed "faculty") gave their own introductory lectures over the following weeks.

Almost from the beginning, Bell was dissatisfied with his lot at the university and constantly threatened to resign. In his correspondence with the warden and the council, he referred frequently to his title and that of Pattison, issues with the granting of certificates,[17,28] the inability of the university to grant degrees (limiting its effectiveness),[28] his desire for more space to house an anatomical museum and a curator to manage the collection,[29,30] and his need to draw on personal funds to conduct the university's business.[31] He warned also about excessive informality developing between students and professors—for example, when a number of the students wanted to take their professors out to dinner, Bell feared it "would lead to a relaxation of discipline."[32]

Early in 1830, it was announced that the young university—unable to confer degrees—would award the diploma of Master of Medicine and Surgery in the University of London (to be abbreviated M Med et Chir U L). Candidates would be required to complete its courses and satisfy a number of other conditions related to level of performance, attendance, ability to translate from Latin to English, the authoring and defending of an essay on an approved topic, and the passing of examinations on any part of their professional studies that the faculty might choose to question.[33] In selecting the designation, care was taken to avoid any interference with "the titles and privileges conferred by chartered bodies," by which was meant the established universities and professional colleges or societies.

From the first, the students did not hesitate to air their grievances—and there were many—in public. One specifically concerned Bell's teaching style. In February 1830, a "senior pupil" wrote to the *Lancet* complaining that Bell's lectures were very deficient in "practical information." A number of cases had been discussed—such as fractures and head injuries—but without any mention of their treatment. Bell himself had explained his approach in an earlier introductory lecture:

> It is an easy matter to name a disease, its symptoms; relate a case, and give the treatment in a "regular routine," as other lecturers do; but I shall attempt a more arduous task in

explaining to you the "principles" on which to act, and these are to be sought for in morbid anatomy.

The students, however, wanted Bell to lay down both the principles involved and also the "regular routine" to which he referred, and they pointed out that his failure to do so accounted for the poorer attendance at his lectures on surgery than on physiology. Perhaps there was some justification in their complaint, but perhaps, also, they simply did not like to think for themselves. In any event, the *Lancet* could not resist yet another opportunity to mock Bell's Scottish burr, titling the published letter "Spicimin" of "lictiring."[34] On another occasion, the editorialist at the journal became even more offensive:

> Mr. Bell's efforts as a lecturer are so oppressed and obscured by an unsightly, sickening affectation and mannerism, that he scarcely ever succeeds in fixing the attention of the student to the subject under consideration; he fatigues more frequently than he instructs, annoys more frequently than pleases.[35]

Not all the students felt this way, however, and a "junior pupil" wrote on behalf of all the class to commend his teaching.[36]

One of the first major problems at the university related to the anatomy department and in particular to the differing responsibilities and interactions of its academic staff and their relationships with the administration (Fig. 11.2). The issues centered—in particular—around Granville Sharp Pattison, a member since 1813 of the faculty of physicians and surgeons of Glasgow, who was charged (but acquitted) in a body-snatching case in Scotland and who subsequently moved to the United States after an alleged affair with the wife of one of his colleagues had involved him in divorce proceedings. Self-opinionated and argumentative, he returned to London from America in 1827 after professional disputes and a duel, and was appointed professor of anatomy at the new university and surgeon to its dispensary. Unfortunately, the university council appointed as his demonstrator someone of its own choosing (James Richard Bennett) rather than allowing the professor to select his own junior colleague, and this led to increasing friction between the two anatomists over financial, administrative, and academic matters.[37] The able demonstrator was independent of the professor, was able to give his own lectures as well as teach in the dissection rooms, and was very popular with the students. Bell and John Conolly tried to settle the issues, but then Bell and Pattison had a falling out.

Pattison believed that Bell had always wanted the chair in anatomy. In 1829, Bell supposedly agreed, without Pattison's knowledge, to a suggestion by Horner that he take over the chair. The idea, which apparently originated with Bennett,

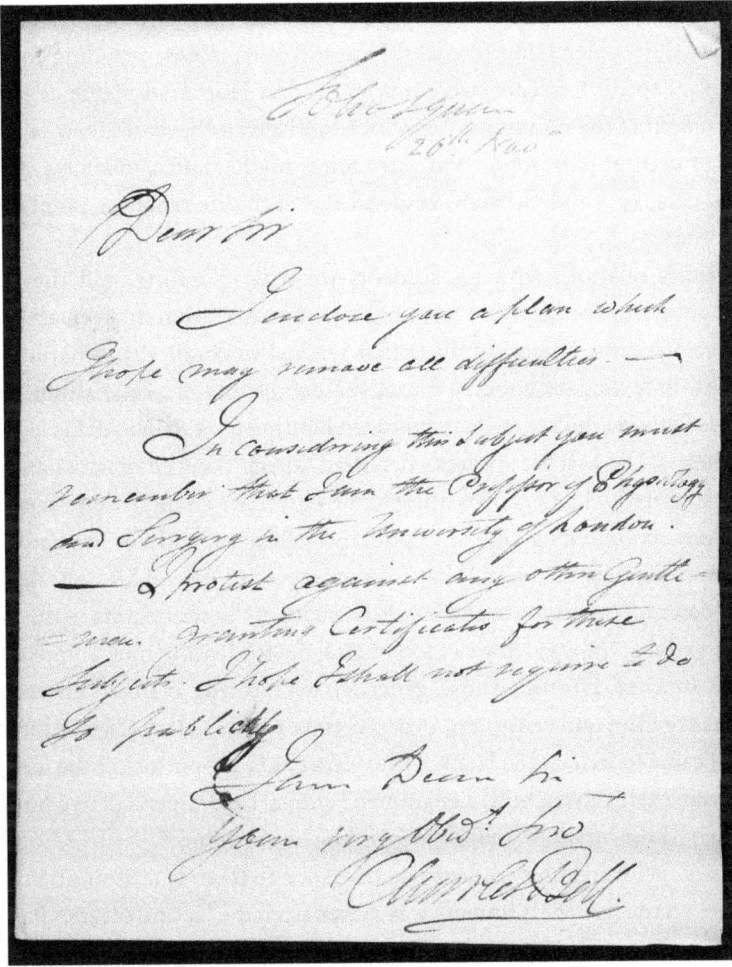

Figure 11.2 Letter dated 20 November (1828) from Bell to Leonard Horner (warden of the University of London) regarding the arguments between the different professors in the medical department concerning their responsibilities and privileges. The body of the letter reads: "I enclose you a plan which I hope may remove all difficulties. In considering this subject you must remember that I am the Professor of Physiology and Surgery in the University of London. I protest against any other gentleman granting Certificates for these subjects. I hope I shall not require to do so publicly." (Courtesy of UCL College Archives; College Correspondence, item 605, UCL Library Special Collections.)

was rejected by the council. A few weeks later, Bell complained to the council that Pattison was representing himself as professor of anatomy and of operative surgery on his class tickets,[28] but Pattison justified his action and refused to alter the tickets.[J] There were further complaints by Bell that Pattison was

---

[J] Tickets were required to attend classes or lectures and had to be purchased.

encroaching on his subjects, Horner moved to support Bell, and Pattison complained of the warden's interference. These and other issues, including copies of the relevant correspondence, were summarized by Horner in a letter addressed to the council of the university,[38] to which Pattison and several others (the professors of natural philosophy and astronomy, mathematics, medicine, Greek, political economy, and German) responded,[39] each side rebutting points made by the other.

Pattison's relations with his students were deteriorating, and they complained about his incompetent teaching style and his failure to keep abreast of new developments, especially those that seemed to conflict with natural theology. An independent enquiry found the complaints baseless, although the eccentric Pattison did sometimes appear in hunting pink to deliver his lectures. Nevertheless, the activist students persisted, airing their complaints and personal insults in the *Lancet* and writing to the university council[40]; one described his lectures as "one long course of puff and nonsense, and not of anatomy."[41] Other students defended him.[42] Pattison attempted to defend himself, showing the self-contradictory nature and inconsistencies of the complaints against him, and charged that the warden was encouraging the students in their complaints.[43] He also attacked Horner, whose generally overbearing manner, attempts to micromanage the university, and overgenerous annual salary (twelve hundred pounds) came to irritate the faculty, who were paid a proportion of the fees generated from their classes, with a guaranteed annual base stipend of two hundred and fifty or three hundred pounds, depending on the subject.

Bennett, beloved by his students, died in April 1831, and Pattison, attempting to lecture on the day of his funeral, was driven from the lecture theater. Pattison was dismissed from his posts later in 1831,[44] after three years in office, when certain students and colleagues continued to question his competence, and he eventually returned to America. Several of the professors in other departments immediately resigned to protest his dismissal. Pattison himself wrote to the *Lancet* complaining bitterly—and with some justification—about the behavior of its editor:

> No man doubts your critical acumen. No man denies your talent for tearing to tatters and exposing to scorn any publication to which you feel hostile, provided *only* that that publication has a single weakness or inconsistency on which you can fasten the sharp fangs of your criticism.[45]

Bitterness and support both for and against Pattison persisted for years. There were lasting consequences—the administration of the university was changed, the office of the warden was abolished, an executive committee of the council was set up, the regulations governing tenure and powers of the professors were formalized, an academic senate was established, and attempts were made to improve the interactions of the academic and administrative personnel.[37]

By this time, Bell had left the university. Increasingly unhappy with his position, his heavy responsibilities, and the acrimonious environment, in September 1830 he gave up the chair of surgery (to be succeeded by—of all people—Pattison, who was dismissed soon afterward) and of clinical surgery (which was left vacant until 1834, when it was filled by Robert Liston).[46] He notified Horner privately toward the end of October that his current series of lectures on physiology would be his last.[47] Without Bell's permission, Horner passed this information on to the council, which immediately started the search for a replacement as if he had already resigned. Bell thereupon refused to finish the course that he was giving on design as exhibited in animal structure, and he left forthwith.[17] He felt betrayed both personally and professionally by Horner. As he stated in a letter to Lord Auckland (a member of the university council), his "resignation had been accepted before it was offered."[48] He felt humiliated and bitterly disappointed that, as the only lecturer from the old private schools to be taken on by the university, his experience and advice had been ignored and one of his most sincere aspirations—to establish a school focused on intelligent design—had got nowhere. His reasons for leaving were many. He spelled them out in part in a letter to his students, dated 25 November 1830, and published by the *London Medical Gazette* in 1831.[49]

He objected, he wrote, not to the conduct of individuals or to the qualifications of his associates but, rather, to the fact that the council had no experience of running a medical school. He had agreed to accept the title of professor of physiology—a science for which he had little regard—in the hope of seeing to success the new educational enterprise, but his lectures were on "the higher departments of anatomy." He had thought that he was to preside over the anatomy department, with the subject organized systematically so that the various professors could each display their various talents. Instead, appointments were made by the council before any such organization could occur, and individual faculty members were not selected to complement each other. Apparently, at the time of his writing, no fewer than five gentlemen were engaged in teaching human anatomy, with three "lecturing in the same classroom, on the same subjects, and with the same preparations put upon the table, three successive times in the same day." He was increasingly disappointed that his designation and duties had not been clarified despite his having requested this within a few days of the opening of the university. He believed that the council was constantly impugning him: He had suggested, for example, that to improve their educational experience, students at the Middlesex and St. George's Hospitals should be able to attend the clinical courses at either hospital, and this was held to imply he had withdrawn his support from the class of surgery at the university.[49]

The editors of the *London Medical Gazette* regarded Bell's letter as simply corroborating their own opinion:

> One of the radical and most conspicuous blunders in the London University consisted in trusting the arrangements regarding the medical school to those alike ignorant of the science of medicine and unacquainted with details. . . . The governing body is made up almost entirely of lawyers and merchants, nor would it be easy to select a class of men less qualified by the nature of their pursuits and occupations to regulate the business of a medical school.[50]

It all became too much for Horner who, in 1831, resigned, in part because of ill health. As he pondered how to announce his resignation to council, Horner asked Bell for his advice and was told, "Stand up, show yourself, and say, 'Gentlemen, I came to your university comfortable and well filled up—look at me now, shrivelled and thin, my clothes a world too wide.' That would be true eloquence."[37]

It was not only Bell, Bennett, and Pattison who left the medical department. John Conolly (1794–1866), professor of the nature and treatment of disease, resigned his chair in 1830. Conolly—a man of culture and charm, formerly in private practice and one of the founders of what was to become the British Medical Association—had been unable to continue for financial reasons because the guaranteed initial salaries dwindled as money became more scarce. After almost four years, he left London for the provinces and became an authority on mental illness. He pioneered the treatment of patients in mental asylums without any forms of mechanical restraint.

From the onset, it had been recognized that the medical school would need an affiliated hospital to provide clinical training, and this need increased once relations with the nearby Middlesex Hospital broke down. A shortage of money was one reason that a hospital had not been built. By October 1832, the university was in debt by almost three thousand pounds, and its council was being accused of mismanagement.[51] Nevertheless—as pointed out acerbically by the *Lancet*—this was no reason to delay.[35] The money for the hospital would come by public subscription and medical student fees (at less than one-fourth the rate paid to other London hospitals), and running costs would be minor:

> If an hospital have only one hundred beds, and if, as is the universal practice (and a very proper one it is), *one bed* be permitted to accommodate *one patient*, it requires not the genius of a Newton to tell us that the hospital can contain no more than one hundred patients at any given period.[35]

Based on full occupancy, the editorial staff calculated it would cost two shillings per day to keep a patient in the hospital, or fourteen shillings per week, thus costing less than four thousand pounds for the full one-hundred patients. To justify these figures, which they considered generous, the editorial staff pointed out that agriculture workers averaged twelve shillings weekly and weavers in the north of England receive no more than seven shillings a week—and on such sums, they had to maintain themselves and their families. In any event, in 1834, the university—which initially had only a dispensary so that students not attending a hospital could witness the actual treatment of disease[52]—opened its own hospital (originally the North London Hospital but later renamed University College Hospital). It contained one hundred and thirty beds, but when completed it was intended to house another one hundred patients. Interestingly, the first person listed as appointed to its consultant staff was Bell—in 1828, before the hospital was built but at a time when the need for it was already recognized—and he had already resigned.

There was continuing opposition to granting the original University of London a degree-conferring charter.[K] With the subsequent opening of King's College London as a rival institution, the original university was renamed University College London (in 1836) and became part of a new, separate University of London that was established by Royal Charter, with the power of conducting examinations and conferring degrees on students from these two colleges (and subsequently from other approved institutions).[L]

What, then, should be made of Bell's interactions at the university? Bell had held his chairs only briefly and believed his connection with the new institution was very unsatisfactory. He had come to the university with an enormous reputation as a scientist, surgeon, clinician, and teacher, and his appointment therefore brought prestige and luster to the new institution as it struggled to find its feet. At the same time, the early bureaucratic problems that occurred at the new school were a constant irritation to him, for he believed passionately in the correctness of his own views, was impatient with and intolerant of the views of others, and found it difficult to accept compromise. He came to regard every disagreement with his views as a personal challenge. He was older than many of the other professors, had reached the apex of his career, and seemed to feel that his seniority granted him special privileges. His early belief—arising perhaps because of miscommunication, a

---

[K] The opposition was from the Universities of Oxford and Cambridge, the Colleges of Physicians and Surgeons, and some of the medical schools and teachers in London.

[L] The University of London is now a federation with a number of constituent colleges. The oldest are University College London and King's College London. Among the others are Birkbeck, Goldsmiths, Queen Mary, the Royal Holloway, and the London School of Economics. For most practical purposes, the constituent colleges function as individual universities.

misunderstanding, or his being deliberately misled—that he was to preside over a large and important department led to disappointment when it became clear that this was not the case. He needed to be in charge, as he had been at Great Windmill Street, and found it impossible to accept that the new medical school was under the control of a committee that had no knowledge of the special requirements for providing a good medical education.

On an intellectual level, Bell must have been hurt that his beliefs were regarded as outdated by many of his colleagues, such as Bennett and Grant, who favored the views of the French anatomists and the emerging concepts of evolution. Regardless, he framed many of his own lectures around the concept of intelligent design (see Chapter 10), as he had from the very beginning of his time at the university. The quarrels that developed at the university among the students and between the students and professors were bad enough. But the personal animosity that developed between the various professors, their ill-defined responsibilities, which led to quarrels and turf wars, and the increasing distrust between the professorial and administrative staff—all exemplified by the Pattison affair—conspired to drive away several talented people from the new institution, including but not limited to Bell.

THE MIDDLESEX HOSPITAL MEDICAL SCHOOL

Many of the students at the Middlesex had come from the Great Windmill Street School and then from the University of London, but with the closure of the former and the opening of its own hospital by the latter, it became clear that the Middlesex Hospital would need to build its own medical school to continue as a teaching institution. Hospitals gained in prestige and importance in the medical community from their association with a medical school. There were at the time more than two hundred beds accommodating some sixteen hundred inpatients each year at the Middlesex Hospital, which was the only hospital of its size in London without its own medical school. Not surprisingly, then, its medical officers requested early in 1835 that the board of governors establish a complete medical school of its own, affiliated with the hospital and under its regulation. Such a development would facilitate the selection and appointment of good junior staff and—it was admitted—provide increased compensation to the staff of the hospital. The submission was signed by Bell and the physician Thomas Watson, as well as by four others including Herbert Mayo. The governors concurred.

In a letter to George in June, Bell wrote,

> We have founded a school in the garden of the Middlesex Hospital. The building will be a complete little thing—theatre, museum, clinical class-room, and dissecting room. . . . I promise to the extent of sixty lectures. To the work I have no objection, but there will be a

great outlay also, although, from the way in which it is taken up by our governors, I believe subscriptions will cover all expenses. The building will cost 2400 pounds.[53]

For once, things moved speedily, and in October 1835 the new medical school was opened. Bell gave an introductory address at its opening, explaining the difficulties both the hospital and he personally had experienced with the university, sparing no detail. He also spoke of the evidence he had given to a parliamentary committee set up to examine medical education (discussed later), of the inequities that he perceived to exist in the way hard work and merit were rewarded by the existing system, and of the fact that some of the greatest men had died poor and even pilloried for their scientific attainments.[54] He seemed to be speaking—sadly—about his own career.

Bell wrote to George just a week later: "I have delivered six lectures, such as only long experience and study could have produced. I lecture to some sixty pupils—which, for a beginning, is as much as we could expect." He added, "I have received not one guinea from these lectures, and expect none. On the contrary, I have subscribed 50 pounds as one of the hospital surgeons, 30 pounds as a lecturer."[55]

The new medical school flourished, eventually becoming one of the leading schools in the capital, but it did so without Bell, who had been so instrumental in establishing it. He accepted the chair of surgery in Edinburgh and was gone by the end of the year. With the reorganization of medical schools that occurred in London in the last quarter of the twentieth century, the Middlesex and University College Hospitals merged their medical schools in 1987, and the combined school subsequently merged with that of the Royal Free Hospital to form an enterprise that in 2008 was renamed the University College London (UCL) Medical School. It is one of the finest medical schools in the world.

EDINBURGH UNIVERSITY

Within a few weeks of the opening of the Middlesex Hospital Medical School, Bell began to consider seriously the possibility of returning to Scotland to accept the chair of surgery in the University of Edinburgh, which had been vacated by the death of John Turner, its previous incumbent. Bell had been selected after it became clear that Robert Liston (1794–1847), an Edinburgh-trained surgeon now living in London, did not want to be considered for the position. Ironically, Liston had been appointed in 1834 to the chair of clinical surgery previously occupied by Bell at University College London, and he chose to remain there.[M]

[M] Famed for both his rudeness and his surgical skill, in 1846 Liston performed the first operation in Europe under ether anesthesia at University College Hospital, London.

The members of the town council of Edinburgh then decided unanimously to ignore various solicitations from local surgeons and all other candidates, and they invited Bell to return to their city.[56] He wrote to George, "But I do not think that you or they know what I shall have to resign. My hospital, my place in the Council of the College, my honourable professorship, my ornamented home, my practice, and my attached friends."[57] Nevertheless, by mid-December he had accepted the position. The idea of returning to Edinburgh and his old university, to the city in which he had grown up, was difficult to reject. In August 1836, he wrote to his niece from London:

> The house is in a bustle. Books gone—pictures packing! People surveying the house! This does look like a change. All my sacred corners usurped—a naked house—no longer a home.
>
> I leave no enemy behind me, and Marion is universally beloved. . . . "Why, then," as they say, " go?" Because there is a time, and that time draws near, when London is intolerable. Every friend away: These streets and squares deserted. . . .
>
> Without independent fortune, the relations which we have formed with society are not without their drawbacks. I must be independent, and through exertion more than fortune. I must pursue that course through which I have attained station, to feel comfortable. I could have made a fortune, and so my friends say, but I could not also attain to what I am, and to what they would have me.[58]

The *Lancet* noted, "The only thing we have to regret in making this announcement [of Bell's appointment to the Edinburgh chair] is, that the emoluments of the office are not commensurate with the value of those services which Sir Charles Bell has already conferred on mankind" (Fig. 11.3).[56]

OTHER EFFORTS AT MEDICAL EDUCATION AND HEALTH CARE REFORM

In 1824, while still the youngest surgeon at the Middlesex Hospital, Bell had written an open letter (published in the *Lancet*) to the governors of the hospital regarding the appointment of a new physician to succeed a senior man.[59] He pleaded for someone experienced and efficient, advocating that the "situation of Physician or Surgeon to an hospital should be a reward for professional merit." In doing so, he argued cogently against the common contemporary practice of appointing junior members of the College of Physicians simply so that they could "learn their profession, and be prepared for private practice," pointing out that this excluded many able men and was unfair to both the patients and the medical profession. And he expressed his concerns about those focused on the rewards of private practice: "[I]t is humiliating to see emolument so exclusively

Chapter 11. New Classrooms: Old Struggles

Figure 11.3 Portrait of Charles Bell, oil. (Courtesy of Wellcome Library, London.)

the motive to action, and it is a life calculated to dissipate a mind even of the highest order." Clearly, Bell was looking to the development of a meritocracy, but this would be years in the making. The *Lancet* responded bitingly that his letter was intended to bias the governors in their selection and that it was "arrogant and hypocritical in the extreme, and wants common candour and honesty to recommend it."[60] Thomas Wakley, it seems, liked to run with the hare and hunt with the hounds.

For several years before he left London, Bell became increasingly active as a member of the council of the Royal College of Surgeons (1830–1836), where he clearly hoped to put reforms into place. Indeed, in 1832, he served on a committee, chaired by Sir Astley Cooper, "to consider the present state of the College."[61] The committee met eighteen times before reporting to council in January 1833, with recommendations for reducing the study time required for the award of the college diploma, establishing a more advanced surgical examination, developing a diploma in midwifery, and setting up a number of new committees.

But other efforts at reform were also in progress. An unintended insult to naval surgeons that arose inadvertently at the Admiralty led to a bitter dispute between Wakley and the leaders of the College of Surgeons, who ejected him from one of its council meetings and started legal proceedings against him and others. Perhaps in consequence, on 16 March 1831, Wakley called a public meeting at the Crown and Anchor tavern in the Strand, London, to establish a national medical institution giving equal rights and equal titles to its fellows. It would be called the London College of Medicine and would subsume the functions of the Royal Colleges of Physicians and Surgeons and the Society of Apothecaries. All qualified practitioners—physicians, surgeons, and apothecaries, whether in private or in hospital-based practice, whether teachers at universities, hospitals, or private schools—would be eligible for fellowship of the college without religious discrimination, and after public examination they would receive a diploma at low cost (three guineas for practitioners; five guineas for students) and bear the title of doctor. Its officers and governing body were to be elected annually. Nepotism would be replaced by public competition in making hospital appointments, and the need for course certificates from approved schools would be removed, allowing students to choose the teaching courses they attended based on their merits. The combined opposition of the existing professional societies and colleges eventually throttled the project, but not the realities of the inequities and problems that were aired. And Wakley came to understand he would have to persuade Westminster if reforms were to occur.

In parliament, Henry Warburton—encouraged by Wakley—asked for a select committee to examine the various aspects of the training and practice of the different branches of the medical profession in the United Kingdom.[N] Such a committee was duly appointed, with Warburton as its chair. It met for several months in 1834, taking evidence from a number of prominent medical and surgical personalities including Bell. Much of the questioning focused on the divisions within the profession (physicians, surgeons, and apothecaries) and their claims for special rights, but some witnesses, especially Bell, worried particularly about educational standards. He gave his views about the structure of the College of Surgeons and of its council; the type of teaching (lectures, lecture–demonstrations, and bedside teaching), duration of training (three years full-time, if not more), and syllabus he believed was required for surgeons; the educational prerequisites (classics, mathematics, and natural philosophy) for commencing professional training; the need to allow qualified surgeons to practice anywhere in the United Kingdom; and about many other issues.[62] He concluded that the total expense for a student

---

[N] Henry Warburton (1784–1858), English politician, was a supporter of Henry Brougham in founding London University and served on its first council. In addition to serving as chairman of a parliamentary committee on the medical profession, he introduced the Anatomy Bill (passed in 1832) concerning the use of cadavers for medical research and teaching, despite much opposition and hostility.

to undergo three years of training (including living expenses) along the lines he suggested would be five thousand nine hundred and nine pounds.

Nothing seems to have resulted from the meetings of the committee, and it was not until 1858, with the passage in parliament of the Medical Act, that the condition of medical education in the United Kingdom began to change. Bell later wrote to George,

> Mr. Warburton, the chairman of the committee, was pleased to say that he had got more from me than he had altogether. I returned the compliment by saying that he used the probe with great dexterity. Warburton's perseverance and acuteness are remarkable. We exhausted each other. For my part I was bathed in perspiration, and my temples beating from this keen encounter.[63]

REFERENCES

1. Rivett G: *The Development of the London Hospital System 1823–1982*. p. 37. King's Fund Publishing Office: London, 1986.
2. Power D'A: The rise and fall of the private medical schools in London. *Br Med J* 1895; I: 1388–1391, 1451–1453 (2 parts).
3. Lawrence SC: Entrepreneurs and private enterprise: The development of medical lecturing in London, 1775–1820. *Bull Med Hist* 1988; 62: 171–192.
4. Holloway SWF: The Apothecaries' Act, 1815: A reinterpretation. *Med Hist* 1966; 10: 107–129.
5. Cope Z: The influence of the free dispensaries upon medical education in Britain. *Med Hist* 1969; 13: 29–36.
6. Anon: Regulations of the College of Surgeons, and Apothecaries' Company, relative to the admission of members and licentiates. Apothecaries' Hall: Regulations for the examination of apothecaries. *Lancet* 1825; 5: 18.
7. Godlee RJ: *The Past, Present, and Future of the School for Advanced Medical Studies of University College, London.* pp. 18–19. Bale, Sons, & Danielsson: London, 1907.
8. Anon: Regulations of the College of Surgeons, and Apothecaries' Company, relative to the admission of members and licentiates. Royal College of Surgeons in London. *Lancet* 1825; 5: 17–18.
9. Parliamentary Papers, House of Commons: *Report of the Select Committee of the House of Commons on Medical Education.* H. M. Stationery Office: London; 1834.
10. Anon: Medical, surgical, and anatomical schools. *Lancet* 1825; 5: 18–32.
11. Bellot HH: *University College London 1826–1926.* pp. 36–37. University of London Press: London, 1929.
12. Bell C: Letter to George dated 24 May 1827. pp. 295–296. In *Letters of Sir Charles Bell, K.H., F.R.S. L. & E: Selected from His Correspondence with His Brother George Joseph Bell.* Murray: London, 1870.
13. Merrington WR: *University College Hospital and Its Medical School: A History.* p. 5. Heinemann: London, 1976.

14. Bell C: A representation to the members of the council on the state of the Medical School of the University of London by Mr. Bell. Dated 3 November 1828. Bell correspondence (604), UCL College Archives.
15. Bell C: Letter to George dated 2 January 1828. pp. 299–300. In *Letters of Sir Charles Bell, K.H., F.R.S. L. & E: Selected from His Correspondence with His Brother George Joseph Bell.* Murray: London, 1870.
16. Bell C: Letter to Leonard Horner, Esq., Warden, dated 9 February 1828. Bell correspondence (600), UCL College Archives.
17. Bellot HH: *University College London 1826–1926.* pp. 39–40, 148–149. University of London Press: London, 1929.
18. Council of the University of London: *Second Statement by the Council of the University of London, Explanatory of the Plan of Instruction.* Taylor: London; Longman, Rees, Orme, Brown, and Green: London; Murray: London, 1828.
19. Bell C: Evidence of Sir Charles Bell, 5 May 1834. p. 131, para 5870. In *Minutes of Evidence Taken Before the Select Committee on Medical Education. Parliamentary Papers, House of Commons and Command. Reports from Committees: Fourteen Volumes. Medical Education. Session 4 February—15 August 1834. Volume XIII—Part II.* H. M. Stationery Office: London; 1834.
20. Merrington WR: *University College Hospital and Its Medical School: A History.* p. 7. Heinemann: London, 1976.
21. Bell C: Letter, addressee unstated but probably Leonard Horner, dated 18 August 1831. Bell correspondence (2091), UCL College Archives.
22. Godlee RJ: *The Past, Present, and Future of the School for Advanced Medical Studies of University College, London.* pp. 13–15. Bale, Sons, & Danielsson: London, 1907.
23. Bellot HH: *University College London 1826–1926.* p. 143. University of London Press: London, 1929.
24. Harte N, North J. *The World of University College London 1828–1978.* p. 32. University College London: London, 1978.
25. Anon: Correspondence. *Lancet* 1829; 12: 748–749; *Lancet* 1829; 13: 50; and *Lancet* 1830; 15: 26.
26. Bell C: Letter to George dated 8 July 1829. pp. 303–304. In *Letters of Sir Charles Bell, K.H., F.R.S. L. & E: Selected from His Correspondence with His Brother George Joseph Bell.* Murray: London, 1870.
27. Anon: London University [containing the opening address by Charles Bell]. *Lond Med Gaz* 1828; 2: 565–568.
28. Bell C: Letter to Leonard Horner dated 20 November 1828, together with a hand-written document titled "Present Plan of Lectures on Anatomy, Physiology, and Surgery in the University of London," another titled "Altered Plan of Lectures," with another letter to Horner dated 21 November 1828. Bell correspondence (605), UCL College Archives.
29. Bell C: Letter to Leonard Horner dated 18 September 1827 (item 300); offer of museum, dated August 1828 (item 602); undated letter to Leonard Horner (item 603); letters dated 3 and 4 December 1828, no addressee (items 607 and 608); letters to Leonard Horner dated 17 April 1830 (item P37) and 21 May 1830 (item P38). Bell correspondence, UCL College Archives.
30. Bell C, Pattison GS: Letter dated 20 July 1828 concerning the duties of a curator of the university anatomy museum. Bell correspondence (952), UCL College Archives.

31. Bell C: Letter to Leonard Horner dated 18 September 1827. Bell correspondence (300), UCL College Archives.
32. Bell C: Letter to the council of the University of London dated 12 October 1829. Bell correspondence (1221), UCL College Archives.
33. Anon: University of London: Medical diplomas. *Med Chir Rev* 1830; 13: 204–206.
34. Anon (a senior pupil): "Spicimin" of "lictiring"—Theory without practice. *Lancet* 1830; 13: 664.
35. Anon: The University of London. *Lancet* 1831; 16: 689–695.
36. Anon (a junior pupil): Mr. Professor Bell. *Lancet* 1830; 13: 709–710.
37. Pattison FLM: *Granville Sharp Pattison: Anatomist and Antagonist, 1791–1851.* pp. 150–164, 175–176, 179–181. University of Alabama Press: Tuscaloosa, AL, 1987.
38. Horner L: *Letter to the Council of the University of London.* Moyes: London, 1830.
39. Pattison GS, and others: *Observations on a Letter Addressed by Leonard Horner, Esq. to the Council of the University, Dated June 1, 1830.* Spottiswoode: London, 1830.
40. Anon: Letter. *Lancet* 1830; 13: 897–898; Thomson A: Letter. *Lancet* 1830; 14: 847–848; Eisdell N: Letter. *Lancet* 1831; 16: 763; Anon: *Lancet* 1831: 15: 749–753; and Bree CR: Letter. *Lancet* 1831; 16: 15–18.
41. Anon: Letter. *Lancet* 1830; 13: 897–898.
42. Anon: Letter. *Lancet* 1831; 15: 815–818; Butter G: Letter. *Lancet* 1831; 16: 763–764.
43. Pattison GS: Letter. *Lancet* 1830; 14: 975; Letter. *Lancet* 1831; 16: 793–795; Letter. *Lancet* 1831; 16: 825–829; and Letter. *Lancet* 1831; 17: 82–87.
44. Pattison GS: Letter to the editor of the Lancet. *Lancet* 1831; 16: 573.
45. Pattison GS: Reply of Mr. ex-Professor Pattison to the strictures of the editor of the Lancet. *Lancet* 1831; 17: 209–215.
46. Bell C: Letter to the chairman of the council of the University of London dated 1 September 1830. Bell correspondence (P42), UCL College Archives.
47. Bell C: Letter to Leonard Horner dated 27 October 1830. Bell correspondence (P43), UCL College Archives.
48. Bell C: Letter to Lord Auckland dated November 1830. Bell correspondence (P43), UCL College Archives.
49. Bell C: Mr. Bell's letter to his pupils of the London University: On taking leave of them. *Lond Med Gaz* 1831; 7: 308–311.
50. Anon: London University—Mr. Bell. *Lond Med Gaz* 1831; 7: 305–308.
51. Anon: The London University and the Society for the Diffusing Useful Knowledge. *Athenaeum* 1833; 121–122.
52. Thomson AT: To the proprietors of the University of London. *Lancet* 1831; 16: 748–750.
53. Bell C: Letter to George dated 29 June 1835. p. 341. In *Letters of Sir Charles Bell, K.H., F.R.S. L. & E: Selected from His Correspondence with His Brother George Joseph Bell.* Murray: London, 1870.
54. Anon: Middlesex Hospital school. *Lancet* 1835; 25: 89.
55. Bell C: Letter to George dated 7 October 1835. pp. 343–344. In *Letters of Sir Charles Bell, K.H., F.R.S. L. & E: Selected from His Correspondence with His Brother George Joseph Bell.* Murray: London, 1870.
56. Anon: Editorial. *Lancet* 1835; 25: 470–471.

57. Bell C: Letter to George dated 27 November 1835. pp. 344–345. In *Letters of Sir Charles Bell, K.H., F.R.S. L. & E: Selected from His Correspondence with His Brother George Joseph Bell.* Murray: London, 1870.
58. Bell C: Letter to Miss Bell dated August 1836. pp. 348–349. In *Letters of Sir Charles Bell, K.H., F.R.S. L. & E: Selected from His Correspondence with His Brother George Joseph Bell.* Murray: London, 1870.
59. Bell C: A letter to the Governors of the Middlesex Hospital, from the Junior Surgeon [Charles Bell]. *Lancet* 1824; 3: 274–277.
60. Anon: Comment on Mr. Bell's letter. *Lancet* 1824; 3: 277.
61. Cope Z: *The Royal College of Surgeons of England: A History.* pp. 52–54. Thomas: Springfield, IL, 1959.
62. Bell C: Evidence of Sir Charles Bell, 5 May 1834. pp. 129–138. In *Minutes of Evidence Taken Before the Select Committee on Medical Education. Parliamentary Papers, House of Commons and Command. Reports from Committees: Fourteen Volumes. Medical Education. Session 4 February—15 August 1834. Volume XIII—Part II.* H. M. Stationery Office: London, 1834.
63. Bell C: Letter to George dated 5 May 1834. p. 336. In *Letters of Sir Charles Bell, K.H., F.R.S. L. & E: Selected from His Correspondence with His Brother George Joseph Bell.* Murray: London, 1870.

12

THE EBBING TIDE

In August 1836, Bell and Marion left London for Edinburgh, breaking the journey to stay with friends along the way. Their departure from London was a sad occasion. Bell's professional friends joined together to present him with a silver urn,[A] with an inscription by Sir Henry Halford:[B] "To Sir Charles Bell, K.H., this memorial of their high respect and regard is presented by his Brethren of the Profession on his quitting London 1836."[C] Bell returned to Scotland an old man, his mind less flexible, his creationist beliefs continuing to frame his views as an anatomist and man of science, resistant to emerging doctrines but more tolerant, perhaps, of the beliefs of others. He returned to a more leisurely academic life in an ancient Scottish university than the strife and student activism of London had allowed and also to a more tranquil world of fishing, sketching, and painting (Fig. 12.1). He had lost his sense of urgency, the sense of immediacy that had driven him as a young man, and the sense of purpose that gave a meaning to his endeavors. His youthful ardor and zeal had yielded to a more conservative and restrained approach. The world had moved on and he had been left behind.

The Bells set up home and established a consulting room at 6 Ainslie Place, part of a development of two Georgian crescents placed around a central garden in a fashionable part of Edinburgh. From there it was an easy walk to the center of town, with leafy views of private gardens. Lady Bell wrote,

> We came to Scotland at the right season and were welcomed by all whom we wished to welcome us. . . . The windows of Ainslie Place looked out to the glorious colouring of the north-west skies, to Corstorphine Hill and the distant Grampians. The garden was

---

[A] Bell chose the urn when asked by Sir Charles Locock (1799–1875) what he would like as a parting gift. Locock was Queen Victoria's obstetrician and the man who introduced bromides for the treatment of seizures. Bromides remained the standard treatment until replaced by phenobarbital in 1912 and phenytoin in 1937.

[B] Sir Henry Halford (born Vaughan; 1766–1844) was physician to King George III and to three of the monarchs who succeeded him. He was president of the Royal College of Physicians of London for a remarkable span of twenty-four years.

[C] The silver urn and a cream jug inscribed to Sir Charles Bell by James Dunlop, Jr., are in the Records Office at University College London. I am grateful to Mr. Colin Penman, records manager, for information concerning them.

Figure 12.1 Bell always enjoyed sketching. (Top) Seated man. (Bottom) Hydrophobia. Patients with rabies may become fearful or agitated at the sight of water because of pharyngeal spasms and difficulty in swallowing. Note the glass of water on the left side of the picture. (From Charles Bell papers, 1830 and undated, David M. Rubenstein Rare Book & Manuscript Library, Duke University, Durham, North Carolina.)

in terraces down to the Water of Leith and the walks there among the sweet briar hedges made our home in Edinburgh very delightful.[1]

And for a while, Bell felt very welcome. "We are overwhelmed with kindness. I fight hard against it; but three dinners in the week swallowed, and six refused, is the rate of invitation of the last week" he wrote.[2] Marion seemed to thrive: "Heaven be praised, Marion is stout and well. She looks as if she would survive me thirty years."[3]

His letters, mainly to friends in London, tell of social outings and encounters, generally with professional friends or the titled aristocracy, most of whom were politicians, in whose company he seemed to take particular delight.[D] Astley Cooper visited Edinburgh and this pleased him enormously, as did Cooper's proposal to nominate him for the position of surgeon in Scotland to Victoria, who had become queen in June 1837 with the death of her uncle, King William IV.

And yet, behind the pleasantries, things were not the same. It is often a mistake to return in later life to live in a city that one has left as a young man. Edinburgh had not suffered from industrialization, but it had changed and was not the city Bell had known in his student days. His old haunts and coffee houses were familiar no longer; friends and relatives had aged, developed infirmities, or died; the companions with whom he had struggled against the existing order were now part of the establishment; and the faces at the university were new, even though Monro *tertius* was still in office. He felt, he once said, that on "returning after a long absence to my native city, every remarkable object, every street, every corner, brought to my recollection some circumstance important to life, and it seemed to me as if I walked in a city of tombs."[4]

Financially, he was not well off, and he was disappointed that his reputation did not bring him more referrals and requests for consultation (Fig. 12.2). Even when these came, they were often to provide a neurological opinion rather than to perform a more lucrative surgical procedure. This state of affairs might well have been anticipated given his eminence as a neurologist and the fact that several illustrious surgeons were already well established in the city. He still had his academic duties to occupy him and he enjoyed lecturing; he tried to make his students think, and he taught them to value anatomy especially as a guide to the art of healing. Some of his professorial responsibilities were not always welcome. As he wrote on one occasion, "I have had invitations to the Tweed, which they say is full of fish, but I have been locked up with these young doctors, 165 having offered themselves for examination."[5] He continued to worry about money, as he had for most of his life, for his university chair brought annual compensation of only four hundred pounds:

> I had hoped that my receipts from the University would have enabled me to accomplish a great work on anatomy—a design which was innermost at my heart when coming here.

[D] These included the Scotts of Ancrum, Lord Francis Jeffrey (politician, judge, and editor), Lord Lothian (politician), Lord Cockburn (Scottish judge and historian), the Abercrombys of Airthrey Castle, Lord John Hay (admiral and politician), and Baron Dunfermline (Whig politician and former Speaker of the House of Commons).

A

Lord Belgrave very ill bled three times — is he not 80?
Mr. Cunningham I suppose
succeeds? Your letters watch'd by   Ainslie Place
him —
        Dear John
            This is the end of the week
when my mind has more freedom some cessation from hourly duties. Little present cares.
and when I can think of the inmates of
Fludyer Street. Still I feel here as in a
dream. That one be I might walk down to
my friends in Westminster this ⅚ o'clock! —
    The prevailing disease here is unabated
& when I read the English papers I am much
on your account. I grieved for poor Riddell.
fainting & felt for friend Hope. with his shaken
nerves.   I see many of your friends here
— are invited now & then to pleasant parties.
but I have ↑only↓ dinners also. yours ...

B

Many thanks for taking my good friend Alie for
our representative. Dined in Lady St. Cathcarts yesterday
saw him with him last week. His friend remark that he
is in great spirits & good preservation. I am in hopes of
closing with 6 of your Senators here. — in consultation or two
more !!

                Sir John Richardson Esqr
                Fludyer Street
                Westminster / Lothian

Let me hear from time to time & always first of the
household. Let me not feel that I have lost my surest
friend. Marion has no complaint but looks
poorly as if this country did not agree with her. &
now that I have a side that I shall close this without giving
her the corner I promised — Yours truly your Nell

But I am obliged to attend to business, and at this season parties drop into Edinburgh for consultation at times when I could wish to be in the country, and I never move without having to calculate what I have lost as well as what I have spent; and then I upbraid myself for extravagance.[6]

In fact, he did very little anatomical research in Edinburgh, publishing just three short papers in the *Transactions of the Royal Society of Edinburgh* on the differences between the cranial and spinal nerves, and these do not provide any novel insights.[7-9]

He continued to write, and the lectures on surgery that he gave at the university were published in two volumes as the *Institutes of Surgery* in 1838.[10] It received a mixed review in one of the medical journals, being criticized principally because it included a number of his previously published clinical lectures from the Middlesex Hospital and because of its unbalanced content, with some topics merely outlined but others covered in excessive detail. The reviewer was fussy, occasionally challenged Bell's opinions, sometimes argued incorrectly against specific statements, and at times read implications where none was intended.[11] Some sections of the book were outstanding, but certain of Bell's comments are indeed startling at first glance. For example, Bell writes in his introductory chapter (p. xxiii): "In noticing here a few authorities, I have preferred English authors principally because they stand distinguished by truth of narration." The reviewer quite reasonably commented,

> We deprecate, in the strongest terms, this insinuation as to the good faith of foreigners. No matter what their character or country, they are involved in one sweeping condemnation as persons whose statements are unworthy of credit. What right have we to arrogate to ourselves exclusive purity?

Figure 12.2 (A and B) Letter to John Richardson (1780–1864) from Charles Bell, Ainslie Place (Edinburgh), 3 February 1837 (date from postmark). Bell's wax seal is intact. In the letter, Bell spoke frankly about some of the difficulties he was experiencing in his new situation, including ongoing financial troubles and the inadequacies of his surgical colleagues. The overall tone of the letter is critical and somewhat depressed. (Courtesy of Jeremy Norman's History of Science. com.) The letter begins as follows:

"Dear John, This is the end of the week when my mind has some freedom, some cessation from hourly duties, little present cares when I can think of the inmates of Fludyer Street [where Bell first lived in London]. Still I feel here as in a dream & that awake I might walk down to my friends in Westminster thro G. park!" The last paragraph reads as follows: "Let me hear from time to time & always first of the household. Let me not feel that I have lost my surest friends. Marion [Bell's wife] has no complaint but looks poorly as if this country did not agree with her & now that I have said that I shall close this without giving her the corner I promised [i.e., a corner of a sheet of this letter paper to write a personal note to Richardson]. Very truly yours, C Bell."

Bell's comment can be better understood, however, in light of his past hostility to Magendie, who, he believed, had wrongfully claimed credit for discovering the functions of the different spinal nerve roots (see Chapter 7), and to those French anatomist–surgeons whose evolutionist tendencies were becoming increasingly influential and clashed with his own creationist ideology (see Chapter 10). Again, it is a surprise that Bell, an anatomist, writes on page 7 of his text concerning morbid anatomy that "by pottering in the dead house, we shall not discover the sources of disease." The quote has been taken out of context, however, for Bell continues, "We are informed to what disorder of organs the disease tends, and the source of formidable symptoms, but not the cause from which these mischiefs spring." And it was certainly true, at least when the remark was made, that the underlying pathogenesis—that is, the causes or mechanisms leading to disease—generally remained unknown. In other words, "pottering in the dead house" could reveal, for example, the presence and extent of a cancer but not its cause.

His surgical textbook was followed three years later by *Practical Essays*,[12] two volumes in which Bell discussed several topics of interest to him, including the power of life to sustain surgical operations, the practice of bleeding, the cause of squint, the action of purgatives, the respiratory nerves, the circulation, and spinal disease. The titles of these essays are deceptive; the essay on the action of purgatives, for example, is actually on trigeminal neuralgia and contains a thoughtful account of referred pain and its basis, as discussed in Chapter 9,[13] and the essay on life and surgical operations is directed at the effects of air entering the circulation.[14] Helped by the devout Marion, he also revised or added to his works on natural theology. A new edition of his Bridgewater treatise came out in 1837, and new editions were published for several years even after Bell had died, revised by Marion and her brother Alexander. Marion also helped to bring out a new, expanded version of *Paley's Natural Theology*, containing Bell's supplementary dissertations and a treatise on animal mechanics, but this was not completed until some years later, being published in four volumes in 1851.

In the spring of 1840, the Bells traveled to Europe. After a brief stop in Paris, they went on to Italy, and in the journal that he kept, Bell made watercolor sketches of the enchanting scenery, the ancient palaces, the bell towers and churches, the faithful and the supplicants, and the various colorful characters whom he met. In Rome, where they spent a month, he visited the grave of his brother John. They passed through Genoa, the Vatican, Perugia, Florence, Bologna, Modena, Parma, and Milan, meeting other British tourists, including Lady Davy (widow of the celebrated scientist Humphry Davy) in Rome and Sir Frederick Adam,[E] who showed them around Florence. Everything was exciting,

---

[E] A Scottish general who fought at Waterloo.

and he wrote "I shall enrich my book on Expression beyond belief. Noses, beards—especially beards—lips, throats, demoniacs, and the Lord knows how much—besides the whole art of kissing, and blowing kisses."[15] He wrote to the professors at local universities and medical schools, and they made quite a fuss of him, showing him their prized collections, museums, hospitals, and clinics. On his return home, he incorporated his Italian notes into a new edition of his *Anatomy of Expressions* (which was published soon after his death).

With the death of Astley Cooper in February 1841, Bell felt cut off from his former colleagues in London, and a few months later the death and burial at sea of David Wilkie while on a return voyage from Palestine filled him with gloom. He wrote to Richardson,

> The weather here has been and still is beautiful, but I become uneasy on seeing the sun set so far north. And, good heavens! How the years roll on, dear John, more and more rapidly and fearfully, and we still the same trifling creatures, unsteady in purpose, unsettled in opinion—and see how our old friends depart. Sir Astley's death seems to cut me off from the London profession; now poor Wilkie—and what a death and burial![16]

Then, one of his favorite students, who had dedicated his thesis to Bell, was drowned in a boating accident, increasing Bell's feelings of impermanence. He struggled on, depressed and sad, worried about George who was gradually going blind, "sick of life," but attending to his usual academic duties—lectures, speeches, and the examination of young doctors. Even his clinical practice was unsatisfying: "I get wearied" he wrote to Marion in August 1841, from Cheshire, "exhausted by the sufferings of others. From this I must go to Belfast, and there I shall witness more misery."[17] He too was ill. He had been having attacks or "spasms" of pain for some time, which he attributed to his stomach or "something more"; their precise location is not clear from his letters. The pains gradually changed, however, taking on the characteristics of angina pectoris, often provoked by emotional stress, but it seems that he did not appreciate their true significance.

During the spring of 1842, Bell's symptoms became more troublesome. His financial concerns also increased because of a measure proposed in London that was likely to affect the privileges of Edinburgh University. This was probably the determination by the Poor Law Commissioners for England and Wales[F] that

---

[F] The determination was made by the commissioners in their annual report for 1842 to the Home Secretary (Sir James Graham, a member of Sir Robert Peel's administration). Legislation subsequently removed these restrictions. A report on sanitary conditions was also published in 1842, but no action was taken until a change of government, when a Public Health Act was passed and a Central Board of Health established. This may have been the board to which his friends wished to nominate him.

medical practitioners with Scottish or Irish qualifications were to be excluded by law from acting as paid medical officers of Poor-Law Unions, thereby affecting the utility of qualifications from Edinburgh University. His friends in London suggested that Bell allow his name to be included in a list of nominations to the Board of Health that was being drawn up and that would have guaranteed him an annual income of six hundred pounds.[18] He decided to leave for London when the university session closed to pursue matters further and also, perhaps, to get medical advice about his own health. He reached Manchester, where he had an attack of anginal pain so acute that—as he wrote—he called for death.[19] He continued, "Anxiety of mind brings on my pain, and it has of late come to that degree of severity that I wonder how the animal textures can sustain the tension."[19]

On 27 April, he and Lady Bell reached Hallow Park near Worcester, the residence of a Mrs. Holland whom he and Marion had decided to visit on their way. He seemed perfectly well the next day and was out sketching for much of the time. Indeed, he was "particularly pleased with the village church and some fine trees which are beside it, observing that he should like to repose there when he was gone."[20] That night he felt cold and listless, unable to eat; his anginal pains recurred and became increasingly severe until eight o'clock on the following morning, 29 April, when he suddenly became very quiet. Marion sent for Dr. Carden of Worcester, but by the time he arrived, Bell had died in her arms and appeared to have been dead for some time.[G] "Just four days after this sentiment had been expressed, his mortal remains were accordingly deposited beside the rustic graves which attracted his notice and so recently occupied his pencil."[20]

His grave is in the old churchyard at Hallow. The memorial tablet in the village church bears the words of his friend, Francis Jeffrey:

> Sacred to the memory of Sir Charles Bell, who, after unfolding, with unrivalled sagacity, patience, and success, the wonderful structure of our mortal bodies, esteemed lightly of his greatest discoveries except only as they tended to impress himself and others with a deeper sense of the infinite wisdom and ineffable goodness of the Almighty Creator.
>
> He was born at Edinburgh, 1774; died, while on a visit of friendship, at Hallow Park, in this parish, 1842; and lies buried in the adjoining churchyard.

[G] The date of his death is given by many authors as 28 April 1842, but the account provided by his widow in his published correspondence and the date of death recorded by her and her solicitor-at-law when presenting an inventory of his personal estate for legal purposes (Ref. 21) is 29 April 1842.

Marion duly prepared an inventory of his personal estate in Scotland, which amounted to just less than fourteen hundred and thirty-five pounds sterling, including household furniture, silver plate, and clothes.[21] This was a pitifully small amount for a man of his position and distinction, and it was clear that she would have difficulty managing on her own. The prime minister, Sir Robert Peel, arranged for her to be placed on the civil pension list so that she would receive an annuity of one hundred pounds—a help, certainly, and a sum in keeping with the pension awarded to, for example, the widow of the surgeon Robert Liston, but small compared to the pensions sometimes awarded to the widows of the country's military heroes. As expressed by the anonymous reviewer of a new edition of Bell's essays on expression—published soon after his death—she had received a pension

> out of that most inadequate fund which the parsimony of Parliament has placed at the disposal of the Crown, for the reward, or rather the bare recognition, of the most important benefits which can be rendered to the nation and to humanity.[22]

His beloved brother George survived Bell by little more than a year, dying in September 1843. Marion—his wife for more than thirty years—returned to London to live with her brother, Alexander, and their house became a place for intellectuals to gather. She had many friends, kept up with her husband's former colleagues in the clinical and academic worlds, stayed abreast of the latest developments in science and literature, was a good conversationalist and listener, and remained articulate and clear-minded to the end. She wrote verse occasionally and a few years before her death prepared a new edition of Bell's treatise on the hand.[23] She died in 1876, a few months before her ninetieth birthday, thirty-four years after her husband.

REFERENCES

1. Bell M: Lady Bell's recollections. pp. 420–421. In *Letters of Sir Charles Bell, K.H., F.R.S. L. & E: Selected from His Correspondence with His Brother George Joseph Bell*. Murray: London, 1870.
2. Bell C: Letter to John Richardson dated 26 February 1837. pp. 350–351. In *Letters of Sir Charles Bell, K.H., F.R.S. L. & E: Selected from His Correspondence with His Brother George Joseph Bell*. Murray: London, 1870.
3. Bell C: Letter to John Richardson dated 8 July 1837. pp. 355–356. In *Letters of Sir Charles Bell, K.H., F.R.S. L. & E: Selected from His Correspondence with His Brother George Joseph Bell*. Murray: London, 1870.
4. Bell C: Notes. p. 338. In *Letters of Sir Charles Bell, K.H., F.R.S. L. & E: Selected from His Correspondence with His Brother George Joseph Bell*. Murray: London, 1870.

5. Bell C: Letter to John Richardson dated 28 July 1838. pp. 361–362. In *Letters of Sir Charles Bell, K.H., F.R.S. L. & E: Selected from His Correspondence with His Brother George Joseph Bell*. Murray: London, 1870.
6. Bell C: Letter to John Richardson dated 18 August (?1839). pp. 369–370. In *Letters of Sir Charles Bell, K.H., F.R.S. L. & E: Selected from His Correspondence with His Brother George Joseph Bell*. Murray: London, 1870.
7. Bell C: On the third pair of nerves; Being the first of a series of papers in explanation of the difference in the origins of the nerves of the encephalon, as compared with those which arise from the spinal marrow. *Trans R Soc Edinb* 1838; 14: 224–228.
8. Bell C: Of the origins and compound functions of the facial nerve or portio dura of the seventh nerve; Being the second paper in explanation of the difference between the nerves of the encephalon, as contrasted with the regular series of spinal nerves. *Trans R Soc Edinb* 1838; 14: 229–236.
9. Bell C: Of the fourth and sixth nerves of the brain; Being the concluding paper on the distinctions of the nerves of the encephalon and spinal marrow. *Trans R Soc Edinb* 1838; 14: 237–241.
10. Bell C: *Institutes of Surgery: Arranged in the Order of the Lectures Delivered in the University of Edinburgh*. 2 vols. Black: Edinburgh; Longman, Orme, Brown, Green, and Longmans: London, 1838.
11. Anon: Art. XII—Institutes of Surgery arranged in the Order of the Lectures delivered in the University of Edinburgh. By Sir Charles Bell, K.G.H. FRS L &E, &c—Edinburgh, 1838. Two Vols. 8vo. pp. 353, 380. *Br Foreign Med Rev* 1838; 6; 154–172.
12. Bell C: *Practical Essays* (in 2 parts). Maclachlan and Stewart: Edinburgh, 1841 and 1842.
13. Bell C: Essay IV. On the action of purgative medications on the different portions of the intestinal canal, with a view to remove nervous affections and tic douloureux. pp. 83–104. In *Practical Essays*. Maclachlan and Stewart: Edinburgh, 1841.
14. Bell C: Essay I. On the powers of life to sustain surgical operations. The effects of violence in wounds and in operations—and the causes of sudden death during surgical operations in some remarkable instances. pp. 1–26. In *Practical Essays*. Maclachlan and Stewart: Edinburgh, 1841.
15. Bell C: Letter to Alexander Shaw dated 18 June 1840. pp. 378–380. In *Letters of Sir Charles Bell, K.H., F.R.S. L. & E: Selected from His Correspondence with His Brother George Joseph Bell*. Murray: London, 1870.
16. Bell C: Letter to John Richardson dated 11 June 1841. pp. 388–389. In *Letters of Sir Charles Bell, K.H., F.R.S. L. & E: Selected from His Correspondence with His Brother George Joseph Bell*. Murray: London, 1870.
17. Bell C: Letter to his wife dated 8 August 1841. pp. 391–392. In *Letters of Sir Charles Bell, K.H., F.R.S. L. & E: Selected from His Correspondence with His Brother George Joseph Bell*. Murray: London, 1870.
18. Bell M: Footnote. p. 396. In *Letters of Sir Charles Bell, K.H., F.R.S. L. & E: Selected from His Correspondence with His Brother George Joseph Bell*. Murray: London, 1870.
19. Bell C: Letter to John Richardson dated 24 April 1842. pp. 397–398. In *Letters of Sir Charles Bell, K.H., F.R.S. L. & E: Selected from His Correspondence with His Brother George Joseph Bell*. Murray: London, 1870.
20. Anon: Death of Sir Charles Bell. *Lond Med Gaz* 1842; 2: 265.

21. Bell C: *Testament of Sir Charles Bell.* SC70/1/62. pp. 404–405. Retrieved from http://www.scotlandspeople.gov.uk/content/images/famousscots/fstranscript59.htm; accessed 5 December 2015.
22. Anon (from the British and Foreign Review): Sir Charles Bell's Essays on Expression. *Eclectic Magazine* 1844; 2 (July): 289–297.
23. Anon: Obituary: Lady Bell. *Med Times Gaz* 1876; 2: 584.

13

TO EACH HIS DUE

Charles Bell was born when America was still a British colony and France still had a king. The small monograph on the brain that brought him fame, but not fortune, was printed privately more than two hundred years ago, in the first quarter of the nineteenth century, while Bell was in his late thirties. It was a time when understanding of the nervous system—of its structure and the way that it operated—advanced remarkably and in a manner that made possible all subsequent progress. But to whom was this advance due? And, with hindsight, how should Bell be regarded?

Memories of Bell have faded with time. His was a sober life, not calculated to attract attention or gossip. Many people have never heard of him, and physicians and surgeons know of him primarily because of the facial weakness and its associated signs named after him. When historians have occasion to think of him, it is generally in connection with the controversy related to his claims for priority in making certain discoveries regarding the nervous system—claims that subsequently were shown to be based on tampered evidence, as discussed in Chapter 7. His other contributions are often ignored or forgotten.

It is senseless to speculate how he would be viewed today if these controversies had not arisen. The historical accounting, however, needs some adjustment because the taint of scandalous conduct should not be allowed to cast aside Bell's many other, real achievements. Moreover, the conclusion about his conduct may have been reached unfairly, without any attempt to understand what may have happened or why. The issue of originality is a tangled one that is not as simple as is commonly supposed.

When Bell began his work on the nervous system, the functions of the different parts of the brain were completely unknown. The anatomical complexity of the brain with its divisions into lobes seemed of little consequence, and many believed that the entire brain made up the *sensorium commune*, which received sensory input and somehow generated an appropriate motor output. There was no way of characterizing the nervous system into subsystems that would allow for a better understanding of its operation in health and disease. Bell's recognition that nerves differ in functions—motor, sensory, and "vital" (autonomic)— and in their connections with parts of the brain and spinal cord suggested that the various parts of the central nervous system have different functions. This, in turn, provided a means of categorizing the functions of yet other parts of the nervous system, based on their connections. This concept, now seemingly so obvious, indeed provided a new anatomy of the brain, as Bell suggested in the title of his monograph.[1]

Bell recognized and reported in his 1811 monograph that the spinal nerve roots had different properties, and in the early 1820s, he published detailed accounts of related work in the *Philosophical Transactions of the Royal Society*. He correctly determined that the anterior but not the posterior roots had motor functions, but he did not conclude that the anterior roots were motor and the posterior roots were sensory. This was shown by Magendie after elegant experiments based in part on Bell's approach. With regard to the cranial nerves, Bell and Mayo, initially in collaboration, studied the motor and sensory innervation of the face. The fifth cranial nerve has two roots, one without and the other with an associated ganglion, analogous to the anterior and posterior spinal roots, respectively, whereas the seventh nerve has a single root without a ganglion, similar to an anterior spinal root. It was Mayo who correctly showed that the seventh cranial nerve supplied the muscles of facial expression, while the fifth nerve provided sensation to the face and also innervated (by its non-ganglionated root) the muscles of mastication. The concepts, the initial experimental approach, and any misinterpretations of his earlier studies of the spinal and cranial nerves were due to Bell. The events surrounding this work were detailed in Chapter 7.

There can be no question that Bell made subtle alterations to writings that he originally had published in the early 1820s to give the impression—but without explicitly stating—that he alone was responsible for these discoveries. The motive for changing them therefore requires consideration. Were they altered simply to ensure their accuracy for the convenience of readers, or were they amended to gain Bell the credit that, rightly or wrongly, he believed was his due?

It is conceivable that Bell originally altered his publications simply to improve the accuracy of their content, to ensure that they were up-to-date. If this were indeed the case, however, he should have included an explanatory note to prevent any confusion and certainly should not have used the amended publications to support his claims for priority. It seems more likely that Bell made the alterations in order to claim priority for the specific discoveries that they described. If he falsified the written record deliberately and with the intent of deceiving others, his claim must be regarded as fraudulent. It seems more likely, however, that he regarded his claims as justified. He must have found it particularly difficult to understand—let alone accept—that he had "missed the boat" and that Magendie and Mayo, following through on his ideas and utilizing his experimental approach, had arrived at the definitive conclusions that he claimed as his own—conclusions that at last put some order into the organization of the nervous system. He must surely have recognized this intellectually, at least at some point. But, given his nature, he was able to persuade himself of the truth of his own fiction and thereby came to believe in his own imagined results. As has been said of another public figure, "The truth, in his hands, was swiftly converted from what it was to what it should have been."[2] In this circumstance, he perhaps deserves a little more understanding even if he is not exonerated.

Both Magendie and Mayo tried, at least initially, to credit Bell for his achievements, but Bell would not settle for anything less than totality for the discoveries. He was a complex, ambitious man, always in a hurry and anxious to succeed. When first he came to London as a young man, he rapidly gained the admiration of established, more senior colleagues by his immense talent, willingness to be of service, and personal charm. He was an outstanding anatomist and artist, a watercolorist and draughtsman with unquestionable skill and style. He was hardworking, cultured, dedicated to his work, and talented in its performance, and he tried to project himself as a kind and thoughtful man. Behind this façade, however, the impatience, intolerance, and certainty of his own views shone through as he became more senior, and he could be irritable and difficult. From his published letters—selected by his widow and presumably showing the best side of him—it is difficult to ignore Bell's repeated anxieties about money, the gloom that often enveloped him, the perceived insults to which he reacted strongly, the evidence of his sense of urgency, and the grandiose notions he harbored of his own brilliance, importance, and success.[3] It is foolhardy to make psychiatric diagnoses at a distance, but at least it can be said that a flavor of mild bipolar disease comes through in his correspondence. Anxious for recognition, taking all intellectual challenges as a personal affront, he advanced in his chosen profession with his head held high, but his self-assurance was partly a sham and he needed constant reassurance from his wife, recognition from his peers, and acclaim from the medical and scientific community. As regards the controversy with Magendie and Mayo, he never seemed to have had any doubt that he was in the right.

On the centenary of the publication of Bell's original monograph, Arthur Keith (1866–1955)[A] penned a laudatory address in the *Lancet*, concluding that "on whatever standard one proceeds to judge, Charles Bell must be assigned a first place amongst the world's anatomists. He did for the anatomy of the nervous system what Harvey did for the circulatory system—brought order out of chaos."[4] In response, the physiologist Waller[B] wrote a meticulous but tedious

---

[A] Sir Arthur Keith was a Scottish anatomist, physical anthropologist, and evolutionist who edited the *Journal of Anatomy* for twenty-one years. He first described the sinoatrial node, which is the pacemaker of the heart.

[B] Augustus Desiré Waller (1856–1922) created the first practical machine for recording the human electrocardiogram. He is not to be confused with his father, Augustus Volney Waller, who

rebuttal that reads like a legal brief, in which he concluded harshly that "Bell's claim as a discoverer was a carefully fabricated claim and that the discovery of the distinction between motor and sensory nerves belongs entirely to Magendie."[5] For the next two years, a lengthy correspondence between these two scientists and others was published in the *Lancet*, until the editor felt obliged to call a halt.

One correspondent concluded that "the only matter in which Bell anticipated Magendie was in devising the experiment of dividing the spinal roots. Magendie discovered their functions by doing so, Bell did not."[6] Even if this view is accepted, this constitutes a major contribution by Bell. But Arthur Keith goes further, pointing out that in 1811, there was no reasonable explanation of the organization of the central and peripheral nervous systems. Bell, he said,

> laid hold of a basal fact; he realized that if he could discover the uses of the various parts of the nervous system he could explain the complexity of their arrangement. His merit lies, not in making a reasonable guess as to the function of cerebrum, cerebellum, double nerve-roots, and double nerve-supply, but in having made this guess from his knowledge of human anatomy, he proceeded to test its truth on the bodies of other animals by dissection, and above all by experiment. His reputation as a discoverer does not rest on a quibble as to who discovered the exact function of the nerve-roots, but on the fact that he was the first man that realized that the anatomy of our brain and nerves could be explained.[7]

This seems a just assessment, but even this ignores Bell's other achievements.

Bell's other contributions to neurology, detailed in Chapters 8 and 9, were less controversial but remarkably substantial. They include his views on sensory specificity, phantom pains, referred pain, the sixth or muscle sense, and the phenomenon now known as reciprocal innervation; his discovery of a new nerve (the long thoracic nerve named after him) and new muscle (in the bladder); and his accounts of a variety of clinical phenomena, including the facial palsy and its associated signs, now all named after him, and diverse other clinical disorders of the muscles, spinal cord, and brain. In many instances, such contributions were made not in scientific papers to learned sciences or in academic journals but, rather, in his well-documented lectures, essays, or textbooks. They have, in the main, stood the test of time; they remain as a solid testimonial to his neurological insight and are not easy to disparage.

Bell's accomplishments and style as an artist, draughtsman, and educator left a lasting impression on the fine arts (see Chapter 4) and also provided a graphic representation of the horrors of war in the early nineteenth century

---

described the series of degenerative changes that occurs when a nerve is cut (so-called Wallerian degeneration) and used these changes as a marker to study the course of nerves.

and a manual for military surgeons, both of which remain of historical interest (see Chapter 5). He also played an important role in educating future generations of doctors by his writings and clinical teaching in the hospital as well as by initiating or participating in various educational reforms in both private and university-based medical schools. His aim was to encourage an approach to disease based on an understanding of the underlying pathophysiology and on anatomical principles, to add a scientific underpinning to the art of medical practice (see Chapter 11). His views of anatomy, however, were influenced by his creationist beliefs and, specifically, by the concept of intelligent design that he championed. It was these beliefs that ultimately caused him to be left behind in the swell of evolutionist thought that developed in the nineteenth century (see Chapters 4 and 10).

Finally, Bell was a cool and talented surgeon with considerable technical expertise. Nevertheless, he became anxious before and during operations, with their inherent risks and uncertainties. He went to operations "with the reluctance of one who has to face an unavoidable evil . . .; [his] cheek was often seen to blanch on proceeding to operations performed with the utmost self-possession and skill."[8] As Bell wrote in the preface to his *A System of Operative Surgery*,

> When a surgeon first takes the knife in his hand, and is preparing, with oppressive feelings, to perform an operation, which may terminate the life of his patient, he is not always aware of what is the most difficult to be accomplished. His ideas are vague; his mind not settled to what he is to expect; the circumstances which ought chiefly to engage him are not distinctly before him; and no man has ever performed this painful duty, without feeling that it is in the very course of the operation that he learns what is most necessary for him to know and to practise.[9]

Such sentiments have sometimes been held to support the belief that he was not at heart a surgeon. To the contrary, they exemplify a very common mindset among the best surgeons, in the same way that even the best actors or musicians may experience some degree of pre-performance anxiety ("stage fright"), in the absence of which their performance sometimes lacks luster. Henry Marsh, a modern neurosurgeon, has commented on his own similar anxieties immediately before an operation—beginning with the decision to operate and continuing until, just before surgery, he is often seized by panic, which is swept away once the operation begins.[10]

Charles Bell (Fig. 13.1) has been dead for nearly one hundred and seventy-five years. A record of his immense achievements is necessary so that new generations of scientists and educators, of surgeons and artists, will not forget them, will bear in mind the foundations on which they themselves build, and will not repeat the mistakes made by Bell. His quarrels with others over priority

Figure 13.1 Bust of Sir Charles Bell at the Royal College of Surgeons of England. (© Museums at the Royal College of Surgeons.)

for certain discoveries should be recalled—if at all—with regret for the behavior that marred an otherwise exceptional career. He was a brilliant but flawed human being who contributed much to the advance of knowledge.

REFERENCES

1. Bell C: *Idea of a New Anatomy of the Brain; Submitted for the Observations of His Friends.* Strahan & Preston, 1811. Reprinted in *Medical Classics* 1936; 1: 105–120.
2. Ziegler P: *Mountbatten: The Official Biography.* p. 701. Collins: London, 1985.
3. Bell C: *Letters of Sir Charles Bell, K.H., F.R.S. L. & E: Selected from His Correspondence with His Brother George Joseph Bell.* Murray: London, 1870.

4. Keith A: An address on the position of Sir Charles Bell amongst anatomists. *Lancet* 1911; 177: 290–293.
5. Waller AD: The part played by Sir Charles Bell in the discovery of the functions of motor and sensory nerves (1822). *Sci Progr* 1911; 6: 78–106.
6. Guthrie L: Charles Bell and the motor and sensory functions of the spinal roots. *Lancet* 1911; 177: 1032.
7. Keith A: The Bell–Magendie controversy. *Lancet* 1912; 180: 968.
8. Arnott JM: *The Hunterian Oration Delivered at the Royal College of Surgeons in London on the Fourteenth of February, 1843.* Scott: London, 1843.
9. Bell C: Preface. p. vii. In *A System of Operative Surgery, Founded on the Basis of Anatomy.* Vol. 1. 2nd edition. Longman, Hurst, Rees, Orme, and Brown: London; Cadell and Davies: London, 1814.
10. Marsh H: *Do No Harm: Stories of Life, Death, and Brain Surgery.* St. Martin's: New York, 2014.

APPENDIX 1

Books Authored by Charles Bell

For books of which Charles Bell was not the sole author, the coauthors and order of authorship are indicated. In a number of cases, the title of books changed with successive editions, and in this circumstance the new title is given to prevent any confusion. When the title has not changed, the new edition is simply indented under the title given previously. The details of each volume have been verified individually, except where indicated.

1. *The Anatomy of the Human Body*

    The Anatomy of the Bones, Muscles, & Joints. (Authored by John Bell, illustrated in part by Charles Bell.) Mudie: Edinburgh; Johnson: London, 1793

    Vol. 1. Containing the Anatomy of the Bones, Muscles, and Joints. 2nd edition. Cadell and Davies: London; Mudie and Son: Edinburgh, 1797

        3rd edition. Longman and Rees: London; Cadell and Davies: London, 1802

    Vol. 2. Containing the Anatomy of the Heart and Arteries. (Authored by John Bell, illustrated in part by Charles Bell.) Cadell and Davies: London; Mudie and Son: Edinburgh, 1797

        Vol. 2, 2nd edition, corrected. Longman and Rees: London; Cadell and Davies: London, 1802

    Vol. 3. Containing the Nervous System, with Plates. Part 1. The Anatomy of the Brain, and Description and Course of the Nerves. Part 2. The Anatomy of the Eye and Ear; Of the Nose and Organ of Smelling; Of the Mouth and Organ of Taste; Of the Skin and Sense of Touch. Longman and Rees: London; Cadell and Davies: London, 1802–1803. Authored and illustrated by Charles Bell.

    Vol. 4. Containing the Anatomy of the Viscera of the Abdomen, the Parts in the Male and Female Pelvis, and the Lymphatic System. In 4 Parts, with an Appendix. Longman and Rees: London; Cadell and Davies: London, 1804. Authored and illustrated by Charles Bell.

    The Anatomy of the Human Body. Containing the Anatomy of the Bones, Muscles, Joints, Heart, and Arteries, by John Bell, Surgeon; And That of the Brain and Nerves, the Organs of the Senses, and the Viscera, by Charles Bell, Surgeon. 3rd edition. 3 vols. Longman, Hurst, Rees, Orme, and Brown: London; Cadell and Davies: London, 1811

        4th edition. 3 vols. Longman, Hurst, Rees, Orme, and Brown: London; Cadell and Davies: London, 1816

The Anatomy and Physiology of the Human Body. Containing the Anatomy of the Bones, Muscles, and Joints, and the Heart and Arteries, by John Bell; And the Anatomy and Physiology of the Brain and Nerves, the Organs of the Senses, and the Viscera, by Charles Bell. 5th edition. 3 vols. Longman, Hurst, Rees, Orme, and Brown: London; Cadell: London, 1823

The Anatomy and Physiology of the Human Body. By John and Charles Bell. 6th edition. 3 vols. Longman, Rees, Orme, Brown, and Green: London; Cadell: London, 1826

7th edition. 3 vols. Longman, Rees, Orme, Brown, and Green: London; Cadell: London, 1829

American Editions

The Anatomy of the Human Body, 1st American edition. 4 vols. Collins and Perkins: New York, 1809

2nd American edition. 4 vols. Collins, New York. 1812

The Anatomy and Physiology of the Human Body. Containing the Anatomy of the Bones, Muscles, and Joints, and the Heart and Arteries by John Bell; And the Anatomy and Physiology of the Brain and Nerves, the Organs of the Senses, and the Viscera, by Charles Bell. 3rd American edition. 3 vols. Collins: New York, 1817

4th American edition. 3 vols. Collins: New York, 1822

The Anatomy and Physiology of the Human Body. By John and Charles Bell. 5th American edition. 2 vols. Collins: New York, 1827

6th American edition. 2 vols. Collins: New York, 1834

2. *A System of Dissections, Explaining the Anatomy of the Human Body, the Manner of Displaying the Parts, and Their Varieties in Disease.*

Vol. 1. Mundell and Son: Edinburgh; Mundell: Glasgow, 1798

A System of Dissections: Explaining the Anatomy of the Human Body, the Manner of Displaying the Parts, and Their Varieties in Disease. Volume the First. Containing the Dissections of the Abdomen, Thorax, Pelvis, Thigh, and Leg. 2nd edition. Mundell and Son: Edinburgh; Johnson: London; Longman and Rees: London, 1799–1800 (Published in several parts; not verified)

Appendix to System of Dissections, Part First; Containing Additional Descriptions of the Abdominal Muscles. Mundell and Son: Edinburgh; Johnson: London; Longman and Rees: London, 1800

Vol. 2. *Containing the Dissections of the Arm, of the Neck and Face, of the Nervous System of the Viscera, and of the Brain.* Mundell and Son: Edinburgh; Longman and Rees: London, 1801–1803 (Published in several parts; not verified)

A System of Dissections, Explaining the Anatomy of the Human Body, the Manner of Displaying the Parts and Their Varieties in Disease, 2nd edition. Mundell and Son: Edinburgh; Murray: London; Longman, Hurst, Rees, and Orme: London, 1805 (Not verified)

A System of Dissections, Explaining the Anatomy of the Human Body, with the Manner of Displaying the Parts, the Distinguishing the Natural from the Diseased Appearances, and Pointing out to the Student the Objects Most Worthy of Attention: During a Course of Dissections. 3rd edition. 2 vols. Longman, Hurst, Rees, and Orme: London, 1809 (Not verified)

American Edition

1st American edition. 2 vols. Jefferis: Baltimore, 1814

3. *Engravings of the Arteries, Illustrating the Second Volume of the Anatomy of the Human Body, by J. Bell, Surgeon; And Serving as an Introduction to the Surgery of the Arteries.* Longman and Rees: London; Cadell and Davies: London, 1801

    Engravings of the Arteries; Illustrating the Second Volume of the Anatomy of the Human Body, and Serving as an Introduction to the Surgery of the Arteries. 2nd edition. Longman, Hurst, Rees, and Orme: London; Cadell and Davies: London, 1806 (Not verified)

    3rd edition. Longman, Hurst, Rees, and Orme: London; Cadell and Davies: London, 1810, 1811

    4th edition. Longman, Hurst, Rees, Orme, Brown, and Green: London, 1824 (Not verified)

    American editions

    Engravings of the Arteries; Illustrating the Second Volume of the Anatomy of the Human Body, and Serving as an Introduction to the Surgery of the Arteries. 1st American edition. Finley: Philadelphia, 1812

    2nd American edition. Finley: Philadelphia, 1816

    3rd American edition. Finley: Philadelphia, 1833 (Not verified)

    German editions

    Karl Bell's Darstellung der Arterien zum Unterricht für Ärzte und Wundärzte bei chirurgischen Operationen und insbesondere für diejenigen welche anatomische Prüfungen zu bestehen haben. Nach der dritten Originalausgabe bearbeitet und mit praktischen Bemerkungen begleitet von Heinrich Robbi. Vorrede von Johann Christian Rosenmüller. Baumgartnerschen: Leipzig, 1819 (Not verified)

    Descriptio arteriarum inconibus illustrata: latio donata et in usum studiosae iuventutis accommodate. Baumgärtne: Lipsiae, 1819 (Not verified)

4. *The Anatomy of the Brain, Explained in a Series of Engravings.* Longman and Rees: London; Cadell and Davies: London, 1802

5. *A Series of Engravings, Explaining the Course of the Nerves.* Longman and Rees: London, 1803 (Not verified)

    A Series of Engravings, Explaining the Course of the Nerves. With an Address to Young Physicians on the Study of the Nerves. 2nd edition. Longman, Hurst, Rees, Orme, and Brown: London; Cadell and Davies: London, 1816

American editions

1st American edition. Finley: Philadelphia, 1818

2nd American edition. Finley: Philadelphia, 1834

German edition

Karl Bell's Darstellung der Nerven: zum Unterricht für Ärzte und Wundärzte, bei chirurgischen Operazionen und insbesondere für diejenigen welche anatomische Prüfungen zu bestehen haben. Baumgartnerschen: Leipzig, 1820

6. *Essays on the Anatomy of Expression in Painting.* Longman, Hurst, Rees, and Orme: London, 1806

    Essays on the Anatomy and Philosophy of Expression. 2nd edition. Murray: London, 1824

    The Anatomy and Philosophy of Expression as Connected with the Fine Arts. 3rd edition. Murray: London, 1844

    4th edition. Murray: London, 1847

    5th edition. Bohn: London, 1865

    6th edition. Bohn: London, 1872

    7th edition, revised. Bell and Sons: London, 1877, 1893

    American edition

    Expression: Its Anatomy and Philosophy. With the Original Notes and Illustrations Designed by the Author; And with Additional Illustrations and Notes by the Editor of the "Phrenological Journal." New edition. Wells: New York, 1873

7. *A System of Operative Surgery, Founded on the Basis of Anatomy.* Vol. 1. Longman, Hurst, Rees, and Orme: London; Cadell and Davies: London, 1807; Vol. 2, 1809

    2nd edition. Longman, Hurst, Rees, Orme, and Brown: London; Cadell and Davies: London, 1814

    American editions

    1st American edition. 2 vols. Hale and Hosmer: Hartford, CT, 1812

    2nd American edition. 2 vols. Goodwin and Sons: Hartford, CT, 1816

    German edition

    Carl Bell's System der operativen Chirurgie. Realschulbuchhandlung: Berlin, 1815

    Italian edition

    Sistema di chirurgia operativa, fondato sulla base dell'anatomia di Carlo Bell; tradotto dall'inglese, e corredato di note da Giacomo Barovero. Pomba: Torino, 1817

8. *Letters Concerning the Diseases of the Urethra.* Murray: London; Bell and Bradfute: London, 1810 (Not verified)

   A Treatise on the Diseases of the Urethra, Vesica Urinaria, Prostate, and Rectum. A new edition (with notes by John Shaw). Longman, Hurst, Rees, Orme, and Brown: London; Burgess and Hill: London, 1820

   3rd edition. Longman, Hurst, Rees, Orme, and Brown: London, 1822

   American edition

   Letters Concerning the Diseases of the Urethra. 1st American edition. Wells and Wait: Boston; Edward Earle: Philadelphia, 1811

   German edition

   Abhandlung über die Krankheiten der Harnröhre, der Harnblase, der Vorsteherdrüse und des Mastdarms. Landes-Industrie-Comptoire: Weimar, 1821

   Swedish edition

   Abhandlung über die Krankheiten—afhandling om urinrörets, urinblåsa, prostata och ändtarmens sjukdomar, medkritiska aumärkningar of John Shaw. Öfversärtning frau Tyskau. Rumstedt: Stockholm, 1824 (Not verified)

9. *Idea of a New Anatomy of the Brain—Submitted for the Observations of His Friends.* Strahan and Preston: London, 1811 (Not verified)

   Reprinted in *Baltimore Medical and Philosophical Lycaeum* 1811; 4: 303–318 (The cover date of publication is 1811, but the actual date was 1812.)

   Reprinted in Shaw A: "Idea of a new anatomy of the brain—submitted for the observations of his friends; by Charles Bell, F.R.S.E." To which are added selections from letters written by the author of the essay to his brother, Professor George Joseph Bell, between the years 1807 and 1821. *J Anat Physiol* 1868; 3: 147–182

   Reprinted in *Idee einer neuen Hirnanatomie, 1811. Originaltext und Übersetzung; mit Einleitung herausgegeben von Erich Ebstein. Klassiker der Medizin, msgh. Von Karl Sudhoff.* Barth: Leipzig, 1911 (Not verified)

   Reprinted in Kelly EC: Idea of a new anatomy of the brain—submitted for the observations of his friends; by Charles Bell, F.R.S.E. *Medical Classics* 1936; 1: 105–120

   Reprinted in Gordon-Taylor G, Walls EW: *Sir Charles Bell: His Life and Times.* pp. 218–231. Livingstone: Edinburgh, 1958

   Reprinted in Cranefield PF: *The Way in and the Way out: François Magendie, Charles Bell and the Roots of the Spinal Nerves.* Futura: Mt. Kisco, NY, 1974

   A facsimile was printed privately for the members of the Classics of the Neurology & Neurosurgery Library, 1987

   Reprinted in Appendix 4, this volume

10. *Engravings from Specimens of Morbid Parts, Preserved in the Author's Collection, Now in Windmill Street, and Selected from the Divisions Inscribed Urethra, Vesica, Ren, Morbosa, et Laesa. Containing Specimens of Every Disease Which Is Attended with Change of Structure in These Parts and Exhibiting the Injuries from the Bougie, Catheter, Caustic, Trochar, and Lithotomy Knife, Incautiously Used: With Observations.* Longman, Hurst, Rees, Orme, and Brown: London, 1813

11. *A Dissertation on Gun-shot Wounds.* Longman, Hurst, Rees, Orme, and Brown: London, 1814

12. *Surgical Observations; Being a Quarterly Report of Cases in Surgery; Treated in the Middlesex Hospital, in the Cancer Establishment, and in Private Practice. Embracing an Account of the Anatomical and Pathological Researches in the School of Windmill Street.* 2 vols. Longman, Hurst, Rees, Orme, and Brown: London, 1816–1818

13. *An Essay on the Forces Which Circulate the Blood; Being an Examination of the Difference of the Motions of Fluids in Living and Dead Vessels.* Longman and Co: London; Burgess and Hill. London, 1819

14. *Illustrations of the Great Operations of Surgery, Trepan, Hernia, Amputation, Aneurism, and Lithotomy.* Longman, Hurst, Rees, Orme, and Brown: London, 1821

    German editions

    Erläuterungen der grossen chirurgischen Operationen durch bildliche Darstellung. Aus dem Englischen. Herausgegeben von Carl Gottlob Kühn. Baumgärtner: Leipzig, 1822–1823

    Grundlehren der Chirurgie von Charles Bell. Aus dem Englischen von C. A. Mörer. 4 vols. bevorwortet von C. v. Grafe. Herbig: Berlin, 1838

15. *Observations on Injuries of the Spine and of the Thigh Bone: In Two Lectures, Delivered in the School of Great Windmill Street. The First in Vindication of the Author's Opinions Against the Remarks of Sir Astley Cooper, Bart. The Second on the Late Mr. John Bell's Title to Certain Doctrines Now Advanced by the Same Gentleman.* Tegg: London, 1824

16. *An Exposition of the Natural System of the Nerves of the Human Body. With a Republication of the Papers Delivered to the Royal Society, on the Subject of the Nerves.* Spottiswoode: London, 1824

    Appendix to Papers on Nerves, Republished from the Royal Society's Transactions. Containing Consultations and Cases Illustrative of the Facts Announced in Those Papers. Longman, Rees, Orme, Brown, and Green: London, 1827

    An Exposition of the Natural System of the Nerves of the Human Body. With a Republication of the Papers Delivered to the Royal Society, on the Subject of Nerves. 2nd edition. Longman, Rees, Orme, Brown, and Green: London, 1827 (Not verified)

    The Nervous System of the Human Body. Embracing the Papers Delivered to the Royal Society on the Subject of the Nerves. Longman, Hurst, Rees, Orme, Brown, and Green: London; Taylor: London, 1830

    The Nervous System of the Human Body: As Explained in a Series of Papers Read Before the Royal Society of London. With an Appendix of Cases and Consultation on Nervous

Diseases. 3rd edition. Black, Edinburgh; Longman, Rees, Orme, Brown, Green, and Longman: London, 1836; Renshaw: London, 1844

American editions

An Exposition of the Natural System of Nerves of the Human Body. With a Republication of the Papers Delivered to the Royal Society on the Subject of the Nerves. 1st American edition. Carey and Lea: Philadelphia, 1825

The Nervous System of the Human Body. Embracing the Papers Delivered to the Royal Society on the Subject of the Nerves. 2nd American edition. Green: Washington, DC, 1833

French edition

Exposition du système naturel des nerfs du corps humain, suivie des mémoires sur le même sujet, lus devant la Société Royale de Londres. Merlin: Paris, 1825

German edition

Karl Bell's physiologische und pathologische Untersuchungen des Nervensystems. Aus dem Englischen übersetzt von Moritz Heinrich Romberg. Stuhr: Berlin, 1832 and 1836

17. *The Principles of Surgery, as They Relate to Wounds, Fistulae, Aneurisms, Wounded Arteries, Fractures of the Limbs, Tumors, the Operations of Trepan and Lithotomy. Also of the Duties of the Military and Hospital Surgeon. By John Bell. A New Edition, with Commentaries, and a Critical Enquiry into the Practice of Surgery, by Charles Bell.* Vol. IV. Tegg: London; Griffin: Glasgow; Bell and Bradfute: Edinburgh; Cumming: Dublin, 1826

18. *Animal Mechanics: Or, Proofs of Design in the Animal Frame: The Perfection of Design in the Bones of the Head, Spine, and Chest, Shown by Comparison with Architectural and Mechanical Contrivances.* Society for the Diffusion of Useful Knowledge. Baldwin, Cradock, and Joy: London; Oliver and Boyd: Edinburgh; Robertson and Atkinson: Glasgow; Wakeman: Dublin; Carvill: New York; Wardle: Philadelphia, 1827, 1829 (Not verified)

    Animal Mechanics, by Charles Bell, Jeffries Wyman, Morill Wyman. Riverside Press: Cambridge, MA, 1902

19. *The Hand: Its Mechanism and Vital Endowments as Evincing Design.* Pickering: London, 1833

    2nd edition, Pickering: London, 1833 (Not verified)

    3rd edition, Pickering: London, 1834

    4th edition, Pickering: London, 1837

    5th edition, revised. Murray: London, 1852

    6th edition. Murray: London, 1854

    The Hand; Its Mechanism and Vital Endowments, as Evincing Design. Preceded by an Account of the Author's Discoveries in the Nervous System by Alexander Shaw. 6th edition, revised. Murray: London, 1860

The Hand; Its Mechanism and Vital Endowments, as Evincing Design. Preceded by an Account of the Author's Discoveries on the Nervous System by Alexander Shaw. 7th edition, revised. Bell and Daldy: London, 1865

The Hand; Its Mechanism and Vital Endowments, as Evincing Design, and Illustrating the Power, Wisdom, and Goodness of God. Preceded by an Account of the Author's Researches in the Nervous System, 9th edition [sic]. Bell and Sons: London, 1874

The Hand; Its Mechanism and Vital Endowments, as Evincing Design and Illustrating the Power, Wisdom, and Goodness of God. Preceded by an Account of the Author's Discoveries in the Nervous System, by Alexander Shaw. 8th edition. Bell and Sons: London, 1875. Reprinted 1877, 1885

American editions

The Hand, its Mechanism and Vital Endowments, as Evincing Design. 1st American edition. Carey, Lea, and Blanchard: Philadelphia, 1833

New editions, 1835, 1836

Harper and Bros., New York, 1840; reprinted. 1855, 1864

German edition

Die menschliche Hand und ihre Eigenschaften. von Dr. Hermann Hauff. Verlag von Paul Neff: Stuttgart, 1836

Dutch edition

Beschouwing der menschelijke hand. Uit het Engelsch en Hoogduitsch. Met eene Voorrede van Dr. Quarin Willemier. Van Der Post: Utrecht, 1836 (Not verified)

20. Paley's Natural Theology, with Illustrative Notes, by Henry Lord Brougham, F.R.S. and Sir Charles Bell, K.G.H., F.R.S., L. & E. to Which Are Added Supplementary Dissertations, by Sir Charles Bell. 2 vols. Knight, London; Jackson, New York, 1836

Paley's Natural Theology; With Illustrative Notes, by Henry, Lord Brougham, and Sir C. Bell and an Introductory Discourse of Natural Theology, by Lord Brougham: To Which Are Added, Supplementary Dissertations, and a Treatise on Animal Mechanics by Sir Charles Bell. 4 vols. Cox: London, 1851

American editions

Paley's Natural Theology, with Selections from Illustrative Notes, and the Supplementary Dissertations, of Sir Charles Bell and Lord Brougham. The Whole Newly Arranged, and Edited by Elisha Barrett, MD. 2 vols. Marsh, Capen, Lyon, and Webb: Boston, 1839

Paley's Natural Theology, with Illustrative Notes, & c. by Henry Lord Brougham, and Sir Charles Bell to Which Are Added, Preliminary Observations and Notes. By A. Potter, D.D. 2 vols. Harper and Brothers: New York, 1840. Reprinted 1842, 1873, 1878

21. Institutes of Surgery: Arranged in the Order of the Lectures Delivered in the University of Edinburgh. 2 vols. Black: Edinburgh; Longman, Orme, Brown, Green, and Longmans: London, 1838

American editions

1st American edition. Waldie: Philadelphia, 1840

2nd American edition. Barrington and Haswell: Philadelphia, 1843

22. *Practical Essays.* (In 2 parts.) Maclachlan and Stewart: Edinburgh, 1841 and 1842

German edition

Pracktische Versuche. Ubersetzt von Dr. Bengel. Laupp: Tübingen, 1842

23. *The Organs of the Senses Familiarly Described: Being an Account of the Conformation and Functions of the Eye, Ear, Nose, Tongue, and Skin.* Harvey: London, Unknown date (Not verified)

    A Familiar Treatise on the Five Senses. Washbourne: London, 1841 (Not verified)

24. *A Letter to Members of Parliament for the City of Edinburgh, on Two Bills Now Before Parliament for the City of Edinburgh, for the Improvement of the Medical Profession.* Machlachlan, Stewart: Edinburgh, 1841 (Not verified)

25. *Letters of Sir Charles Bell, K.H., F.R.S.R&E: Selected from His Correspondence with His Brother George Joseph Bell.* Murray: London, 1870

APPENDIX 2

Charles Bell's Published Papers and Lectures

Charles Bell was a prolific author, but most of his work was published in books (listed in Appendix 1). Even those papers that were published in medical or scientific journals were often republished in his books, sometimes with amendments that had major implications regarding claims for priority of scientific discovery. Some of his important original research was published in the *Philosophical Transactions of the Royal Society*. His clinical lectures or review papers generally appeared in the pages of the *London Medical Gazette*, a conservative medical journal founded in 1827 that published clinical reviews and lectures by established leaders in academic medicine and that stood in opposition to the more radical *Lancet* (founded in 1823). Included in this appendix with his published papers are some of Bell's lectures that were transcribed and reported by others, such as his introductory or opening address at the University of London.

Accounts of the muscles of the ureters; and their effects in the irritable states of the bladder. *Med Chir Trans* 1812; 3: 171–190

On the muscularity of the uterus. *Med Chir Trans* 1813; 4: 338–360

On the nerves; giving an account of some experiments on their structure and functions, which lead to a new arrangement of the system. *Philos Trans R Soc Lond* 1821; 111: 398–424

On the nerves which associate the muscles of the chest, in the actions of breathing, speaking, and expression. Being a continuation of the paper on the structure and functions of the nerves. *Philos Trans R Soc Lond* 1822; 112: 284–312

On the motions of the eye, in illustration of the uses of the muscles and nerves of the orbit. Communicated by Sir Humphrey Davy, Bart. P. R.S. *Philos Trans R Soc Lond* 1823; 113: 166–186

Second part of the paper on the nerves of the orbit. Communicated by Sir Humphrey Davy, Bart. Pres. R.S. *Philos Trans R Soc Lond* 1823; 113: 289–307

On the varieties of diseases comprehended under the name of carcinoma mammae. *Med Chir Trans* 1823; 12: 213–234

Letter to the Governors of the Middlesex Hospital, from the junior surgeon. *Lancet* 1824; 5: 274–277

Introductory anatomical lectures: Mr. Charles Bell. *Lancet* 1825; 5: 99–102

On the nervous circle which connects the voluntary muscles with the brain. *Philos Trans R Soc Lond* 1826; 116: 163–173

Observations on fractures of patella. *Lond Med Gaz* 1827; 1: 25–31

Observations on the diseases and accidents to which the hip-joint is liable. *Lond Med Gaz* 1828; 1: 73–79, 137–142 (2 parts)

Of the eyelids: As indicating different affections of the nerves. *Lond Med Gaz* 1828; 1: 110–115

Foreign substances in different parts of the body. *Lond Med Gaz* 1828; 1: 175–176

Observations on the question of amputation. *Lond Med Gaz* 1828; 1: 201–205, 265–268 (2 parts)

Observations on haemorrhage. *Lond Med Gaz* 1828; 1: 361–365, 425–429 (2 parts)

Lectures at the Royal College of Surgeons by Mr. Charles Bell. *Lond Med Gaz* 1828; 1: 460–464

Lectures on the nervous system delivered at the College of Surgeons by Mr. Charles Bell. *Lond Med Gaz* 1828; 1: 553–557, 617–622, 681–686, 745–747 (4 parts)

Clinical lecture on partial paralysis of the face. *Lond Med Gaz* 1828; 1: 747–750, 769–770 (2 parts)

Introductory address on the opening of London University. *Lond Med Gaz* 1828; 2: 566–568

Mr. Bell's clinical observations on the operations upon the urethra. *Lond Med Gaz* 1828; 2: 809–812

Clinical lecture upon hernia. *Lond Med Gaz* 1828; 3: 104–108

Cases of affections of the nerves, with clinical remarks. *Lond Med Gaz* 1829; 3: 337–344

Nervous system [letter to the editor]. *Lond Med Gaz* 1829; 3: 691–692

London University—Mr. Bell's introductory lecture, Oct. 1, 1829. *Lond Med Gaz* 1829; 5: 18–21

Clinical lecture on the operation of laryngotomy. *Lond Med Gaz* 1830; 5: 72–76

Clinical lecture upon cancer of the mamma. *Lond Med Gaz* 1830; 5: 167–171

Clinical lecture on distortion of spine. *Lond Med Gaz* 1830; 5: 232–235

Clinical lecture on diseases of spine. *Lond Med Gaz* 1830; 5: 294–298, 327–333 (2 parts)

Clinical lecture on the operation of puncturing the bladder. *Lond Med Gaz* 1830; 5: 583–587

Clinical lecture upon the operation for hernia. *Lond Med Gaz* 1830; 5: 679–682

On the nerves of the face; Being a second paper on that subject. *Philos Trans R Soc Lond* 1829; 119: 317–330; *Lond Med Gaz* 1830; 5: 455–461

Mr. Bell's letter to his pupils of the London University on taking leave of them. *Lond Med Gaz* 1831; 7: 308–311

Of the organs of the human voice. *Philos Trans R Soc Lond* 1832; 122: 299–320 (abstract in *Edinb Med Surg J* 1833; 39: 491–492; condensed also in *Lond Med Gaz* 1833, 11: 647–654)

Cases of hernia. *Lond Med Gaz* 1832; 10: 742–749

Lectures on the Hunterian preparations in the museum of the Royal College of Surgeons, London. *Lancet* 1833; 21: 279–285, 313–319, 486–492, 912–919, 962–969; and *Lancet* 1834; 22: 216–221, 265–271, 346–352, 410–416, 745–751, 794–806, 824–829, 875–887 (multiple parts)

Clinical lecture upon the larynx and laryngotomy. *Lond Med Gaz* 1833; 11: 280–286

Clinical lecture on diseases of urethra and neck of the bladder. *Lond Med Gaz* 1833; 11: 391–397, 424–430, 489–492 (3 parts)

Clinical lecture on oesophagotomy. *Lond Med Gaz* 1833; 11: 538–541

Clinical lecture on lithotomy. *Lond Med Gaz* 1833; 11: 686–692

Clinical lecture on compound fracture of leg. *Lond Med Gaz* 1834; 13: 298–302

Clinical lecture upon aneurism and the tying of arteries. *Lond Med Gaz* 1834; 13: 329–334, 423–429 (2 parts)

Clinical lecture on amputations. *Lond Med Gaz* 1834; 14: 181–185

Clinical lecture on fracture of the skull. *Lond Med Gaz* 1834; 13: 487–493, 585–589 (2 parts)

Clinical lecture on diseases of the nerves of the head. *Lond Med Gaz* 1834; 13: 697–702

Clinical lecture on diseases of the fifth pair of nerves. *Lond Med Gaz* 1834; 13: 759–767

Clinical lecture on cases of hernia. *Lond Med Gaz* 1834; 13: 921–927, 984–989

Clinical lecture on diseases of hip joint. *Lond Med Gaz* 1834; 14: 296–303

On the functions of some parts of the brain, and on the relations between the brain and nerves of motion and sensation. *Philos Trans R Soc Lond* 1834; 124: 471–483 (reprinted in *Lond Med Gaz* 1834; 15: 614–621)

Clinical lecture on amputation. *Lond Med Gaz* 1834; 15: 90–96

Clinical lecture on wounded arteries of the fore-arm. *Lond Med Gaz* 1834; 15: 202–208

Clinical lecture on stricture, with lacerated urethra and distended bladder. *Lond Med Gaz* 1834; 15: 294–300

Clinical lecture on cancer, more especially on carcinoma of the mamma. *Lond Med Gaz* 1834; 15: 423–429

Clinical lecture on dysphagia. *Lond Med Gaz* 1834; 15: 565–567

Clinical lecture on aneurism. *Lond Med Gaz* 1834; 15: 567–571

Clinical lecture on scrofulous diseases of the hip. *Lond Med Gaz* 1834; 15: 698–704

Clinical lecture on diseases of the knee-joint. *Lond Med Gaz* 1834; 15: 886–892

Decussation of the posterior cerebral columns [letter to the editor]. *Lond Med Gaz* 1835; 15: 513

On the decussation of the posterior columns of the crus cerebri. *Lond Med Gaz* 1835; 15: 626–627

Clinical lecture on diseases of the spine. *Lond Med Gaz* 1836; 17: 231–236

Clinical lecture on compound fracture of the femur. *Lond Med Gaz* 1836; 17: 600–606

Clinical lecture on tic douloureux. *Lond Med Gaz* 1836; 17: 874–877

Clinical lecture on crushing stone in the bladder. *Lond Med Gaz* 1836; 17: 997–1002

On the third pair of nerves, being the first of a series of papers in explanation of the difference in the origins of the nerves of the encephalon, as compared with those which arise from the spinal marrow. *Trans R Soc Edinb* 1838; 14: 224–228

Of the origins and compound functions of the facial nerve or portio dura of the seventh nerve; being the second paper in explanation of the difference between the nerves of the encephalon, as contrasted with the regular series of spinal nerves. *Trans R Soc Edinb* 1838; 14: 229–236

Of the fourth and sixth nerves of the brain; being the concluding paper on the distinctions of the nerves of the encephalon and spinal marrow. *Trans R Soc Edinb* 1838; 14: 237–241

APPENDIX 3

The Drawings and Paintings of Charles Bell

Bell's talent as an illustrator added greatly to his teaching skills and to the quality of his various books, which are generally well illustrated, usually by his own works. Although some of his paintings are lost, this appendix lists the valuable collections that still exist.

THE CORUNNA OIL COLORS AT THE ROYAL COLLEGE OF SURGEONS OF EDINBURGH

The series of fifteen oil paintings of war wounds listed here was made by Bell after the Battle of Corunna and subsequent evacuation (between 24 December 1808 and 11 January 1809) to England, where Bell treated some of the wounded. The paintings are in the museum of the Royal College of Surgeons of Edinburgh. In the absence of contemporary photographs, they are a remarkable representation of the wounded of a Napoleonic battle. They were intended to educate other medical professionals and thus are quite graphic. Several of the illustrations were used by Bell in his textbooks and monographs, in which case descriptions were also given. They are described in detail, with an accompanying commentary on the medical aspects of each case by Crumplin MKH, Starling PH: *A Surgical Artist at War: The Paintings and Sketches of Sir Charles Bell 1809–1815*. Royal College of Surgeons of Edinburgh: Edinburgh, 2005. The pictures can be viewed online at https://www.museum.rcsed.ac.uk, where they can be accessed at the museum catalogue (Collection tab) under *Corunna*.

1. *Long-standing gunshot fracture of the fibula*: A compound fracture of the fibula has been followed by infection, swelling, and ulceration, which eventually necessitated amputation.
2. *Gunshot wound of the clavicle and scapula*: The subject was an officer wounded by a musket ball at Corunna and evacuated to Portsmouth. The missile fractured the right clavicle and lodged in the scapula. An abscess has formed in the front of the chest.
3. *Gunshot fracture of the shaft of the humerus*: The wound and surrounding inflammation are clearly visible.
4. *Gunshot wound of the head*: A spent musket ball has pierced the skin over the forehead and run under the scalp. It remained superficial and was easily removed.
5. *Bullet wound of the skull*: A musket ball has caused a circular depression with fragmentation of part of the skull.
6. *Gunshot wound of the elbow*: A musket ball is in the right elbow joint, which is swollen and inflamed. The arm is wasted both above and below the elbow. Necrotic tissue is sloughing off.

7. *Long-standing gunshot wound of the humerus*: The picture shows the findings two years after the wound was inflicted. There is infection, inflammation, ulceration, and necrosis.
8. *Gunshot wound of the humerus*: The head of the humerus has been shattered by a musket ball, which passed through it and injured a rib. The patient is very pale, either from blood loss, sepsis, or both. The army surgeons proposed to amputate at the shoulder joint, a procedure with which Bell disagreed given the patient's general condition.
9. *Gunshot wound of the testes*: The entry and exit wounds are seen. Both testicles have been injured, and there is marked swelling of the scrotum. The thighs have not been injured.
10. *Bullet wound of the skull*: The musket ball entered above the forehead, penetrated and tracked under the skull, and pushed up fragments of bone that form a lump under the intact scalp. Apparently, the subjacent dura was penetrated, making brain infection and swelling likely.
11. *Opisthotonus, tetanus following gunshot wounds*: This is perhaps the most famous of these fifteen paintings. The body is arched between the head and the heels.
12. *Sketch of a wound of the abdomen*: The patient is bent forward, clutching at his abdomen. It has been wounded by a musket ball, which passed subcutaneously. Both entry and exit wounds, with necrotic sloughing, are shown.
13. *Gunshot entry and exit wounds of chest*: The musket ball entered the front of the chest just to the right of the sternum and passed tangentially and probably superficially to exit near the seventh rib.
14. *A man wounded in the chest*: The musket ball fractured the clavicle, passed through the lungs and chest, and lodged in the scapula. The soldier died in respiratory distress the day after being landed in England. His facial features are indistinct and partly concealed.
15. *Gunshot wound of the thigh*: There is a wound on the inner aspect of the right thigh. There was apparently recurrent hemorrhage from the wound, but the soldier survived.

WATERCOLORS BY CHARLES BELL OF THE BATTLE OF WATERLOO

These watercolors are held at the Army Medical Services Museum, Keogh Barracks, in Aldershot, Surrey, UK. They are also listed in the catalogue of the Wellcome Library, numbered in sequence from RAMC/95/1 to RAMC/95/17, and can be viewed in digitized form on the website of the Wellcome Library. Some extraneous material has been deleted from the list by the present author, and some of Bell's marginal notes around the original drawings in his sketchbook have been added. Further details are provided in Longmore T: *Description of a Series of Watercolour Drawings Executed by the Late Sir Charles Bell Illustrative of Wounds Received at the Battle of Waterloo*. Army Medical School Museum (not dated; also published in Army Medical Department Reports, *Statistical, Sanitary, and Medical Reports Presented to Both Houses of Parliament by Command of Her Majesty*, Vol. 7, pp. 596–605, 1867).

1. *Sabre wound to the head*: A portion of the skull at the vertex is completely detached by the sabre cut, but the scalp remains connected by a small isthmus. There are also facial injuries.
2. *Musket ball wound to head*: The ball entered the cranium in front, passed through part of the brain, lodged behind, and was extracted on the seventeenth day. The patient's

features are compressed, and he also seems to be in some pain. Bell explains this expression in his notebook, indicating that the pain was not from his wound or from the surgeon but, rather, from the hospital barber who was working on him at the time.

3. *Gunshot wound to forehead*: A musket ball has perforated the frontal bone, entered the left orbit, and is causing protrusion of the eyeball on that side.
4. *Gunshot fracture of skull*: A portion of the frontal bone, an inch in diameter, had been driven into the brain and could not easily be extracted. Trepanning was performed. The cross-shaped surgical cut of the skin around the wound, made in order to remove damaged bone fragments, can be seen. The patient died six days later.
5. *Gunshot wound to both eyes* (left figure) *and to the head* (right; entrance wound over left malar bone): There are no histories of these two cases.
6. *Sword wound to the neck*: Soldier with bandaged head and sword (small-diameter) wound to the center of the neck below the cricoid cartilage. "The oesophagus is evidently wounded for almost the whole food passes this way and the saliva keeps trickling." There does not appear to be much local inflammation.
7. *Grapeshot wound to the neck*: Soldier with grapeshot wound to the neck, lying on his back. There is a deep perforating wound of the right lower half of the neck. The "exit of the ball" was slightly above the right acromion. The patient initially had severe hemorrhage. He subsequently died of his wound.
8. *Cannon shot wound to chest*: Cannon shot has caused an extensive superficial wound of the chest, with the flesh of the right breast torn off. There is some staining of the skin around the injury. The head and right arm of the soldier are also injured and have been bandaged.
9. *Sabre wound to the abdomen*: The bowels are protruded and gangrenous.
10. *Gunshot wound to shoulder*: The soldier has a bandage around his head and his right arm in a sling. The head of the right humerus and part of the scapula were shattered, and amputation at the joint was performed because of the extent of the injury.
11. *Gunshot wound to arm*: The head of the left humerus has been shattered by a musket ball that exited behind the bone. Some small pieces of bone were extracted a few days later. This drawing was made approximately five or six days after operation. The shoulder is swollen, red, and inflamed. An abscess is forming. It eventually discharged and healed.
12. *Gunshot wound to arm*: There is a fracture and necrosis of the left humerus.
13. *Cannon shot wound to arm*: The left arm is missing. Part of the clavicle, scapula, and left shoulder have also been carried off by cannon shot. A ligature has been tied around his axillary artery to prevent hemorrhage. The soldier is lying on his side, grasping a rope to help him to maneuver. He eventually did well.
14. *Cannon shot wound to arm*: The right arm has been carried off close to the shoulder joint.
15. *Cannon shot wound to arm*: The right arm has been carried off by cannon shot. Amputation was performed, to remove the head of the humerus from its socket, a procedure that Bell thought was unnecessary.
16. *Arm shattered*: The left arm was shattered from the elbow to the shoulder joint and incomplete amputation had been performed. The patient made a remarkable recovery from an attack of tetanus and survived his wounds. He is holding a quill in his right hand.
17. *Gunshot fracture of the leg*: The fracture is in the lower third of the leg. The swollen limb is supported by a splint made up of a bundle of straw.

UNIVERSITY COLLEGE LONDON ART MUSEUM, LONDON

Bell left a number of his watercolors to the University of London (now University College London; UCL) for teaching purposes. Some are missing, but the following are listed in the UCL Art Museum online catalogue. They do not portray war wounds but, rather, aspects of human anatomy, and they appeared in several of Bell's publications. They can be viewed in digitized form at http://artcat.museums.ucl.ac.uk/brief.aspx (using the search term "Charles Bell").

1. *Side of the face*: Watercolor over graphite and charcoal
2. *Respiratory system*: Watercolor, graphite and iron gall ink
3. *Spinal nerves*: Watercolor, pen and ink
4. *Nervous system of the head and trunk*: Watercolor, pen and ink
5. *Sunday in England*: Landscape; watercolor, pen and ink
6. *Sunday in Scotland*: Landscape, watercolor, pen and ink

DRAWINGS IN THE BRITISH MUSEUM, LONDON

The Department of Prints and Drawings at the British Museum owns twenty-one drawings by Bell. The term "drawings" as used at the museum is a generic one to cover all methods of hand-drawing on paper, including watercolor, pen and ink, and graphite. In its catalogue, the drawings are numbered consecutively from 1887,0312.1 to 1887,0312.21. They are mainly landscapes. They can be viewed in digitized form at http://www.britishmuseum.org/research/collection_online/search.aspx using the search term *drawing*; Sir "Charles Bell".

APPENDIX 4

Idea of a New Anatomy of the Brain

Regarded by some as the *Magna Carta* of neurology, one hundred copies of Bell's *Idea of a New Anatomy of the Brain* were printed privately in 1811 and distributed by Bell to selected friends and colleagues. The original publication was reprinted in the *Baltimore Medical and Philosophical Lycaeum* 1811; 4: 303–318; by his brother-in law and assistant, Alexander Shaw, in the *Journal of Anatomy and Physiology* 1868; 3: 147–182; by Kelly EC in *Medical Classics* 1936; 1: 105–120; and by Gordon-Taylor G, Walls EW in *Sir Charles Bell: His Life and Times*, pp. 218–231. Livingstone: Edinburgh, 1958. A facsimile was also printed privately for the members of the Classics of the Neurology & Neurosurgery Library in 1987. Copies of the original publication are held in the British Museum, the Royal College of Physicians of London, the Royal Society of Medicine, University College London, and the Royal College of Physicians of Edinburgh. Because of its importance and inaccessibility, it is reprinted here in its entirety.

In referring in his *Idea* to the cranial nerves (CN), Bell used the numbering system of Thomas Willis. Their modern equivalents are as follows. The portio dura of CN VII is the facial nerve; the portio mollis of CN VII is actually CN VIII; Bell's VIII nerve is today's CNs IX, X, and XI; Bell's IX nerve is CN XII; and Bell's X nerve is the first cervical nerve.

IDEA OF A NEW ANATOMY OF THE BRAIN;

SUBMITTED

FOR THE OBSERVATIONS OF HIS FRIENDS;

By CHARLES BELL, F.R.S.E.

NOTE.

THE want of any consistent history of the Brain and Nerves, and the dull unmeaning manner which is in use of demonstrating the brain, may authorize any novelty in the manner of treating the subject.

I have found some of my friends so mistaken in their conception of the object of the demonstrations which I have delivered in my lectures, that I wish to vindicate myself at all hazards. They would have it that I am in search of the seat of the soul; but I wish only to investigate the structure of the brain, as we examine the structure of the eye and ear.

It is not more presumptuous to follow the tracts of nervous matter in the brain, and to attempt to discover the course of sensation, than it is to trace the rays of light through the humours of the eye, and to say, that the retina is the seat of vision. Why are we to close the investigation with the discovery of the external organ?

It would have been easy to have given this Essay an imposing splendour, by illustrations and engravings of the parts, but I submit it as a sketch to those who are well able to judge of it in this shape.

THE prevailing doctrine of the anatomical school is, that the whole brain is a common sensorium; that the extremities of the nerves are organized, so that each is fitted to receive a peculiar impression; or that they are distinguished from each other only by delicacy of structure, and by a corresponding delicacy of sensation, that the nerve of the eye, for example, differs from the nerves of touch only in the degree of its sensibility.

It is imagined that impressions, thus differing in kind, are carried along the nerves to the sensorium, and presented to the mind; and that the mind, by the same nerves which receive sensation, sends out the mandate of the will to the moving parts of the body.

It is further imagined, that there is a set of nerves, called vital nerves, which are less strictly connected with the sensorium, or which have upon them knots, cutting off the course of sensation, and thereby excluding the vital motions from the government of the will.

This appears sufficiently simple and consistent, until we begin to examine anatomically the structure of the brain, and the course of the nerves,—then all is confusion: the divisions and subdivisions of the brain, the circuitous course of nerves, their intricate connections, the separation and re-union, are puzzling in the last degree, and are indeed considered as things inscrutable. Thus it is, that he who knows the parts the best, is most in a maze, and he who knows least of anatomy, sees least inconsistency in the commonly received opinion.

In opposition to these opinions, I have to offer reasons for believing, that the cerebrum and cerebellum are different in function as in form; that the parts of the cerebrum have different functions; and that the nerves which we trace in the body are not single nerves possessing various powers, but bundles of different nerves, whose filaments are united for the convenience of distribution, but which are distinct in office, as they are in origin from the brain:

That the external organs of the senses have the matter of the nerves adapted to receive certain impressions, while the corresponding organs of the brain are put in activity by the external excitement: That the idea or perception is according to the part of the brain to which the nerve is attached, and that each organ has a certain limited number of changes to be wrought upon it by the external impression:

That the nerves of sense, the nerves of motion, and the vital nerves, are distinct through their whole course, though they seem sometimes united in one bundle; and that they depend for their attributes on the organs of the brain to which they are severally attached.

The view which I have to present, will serve to shew why there are divisions, and many distinct parts in the brain: why some nerves are simple in their origin and distribution, and others intricate beyond description. It will explain the apparently accidental connection between the twigs of nerves. It will do away the difficulty of conceiving how sensation and volition should be the operation of the same nerve at the same moment. It will shew how a nerve may lose one property, and retain another; and it will give an interest to the labours of the anatomist in tracing the nerves.

IDEA, &c.

WHEN in contemplating the structure of the eye we say, how admirably it is adapted to the laws of light! we use language which implies a partial, and consequently an erroneous view.

And the philosopher takes not a more enlarged survey of nature when he declares how curiously the laws of light are adapted to the constitution of the eye.

This creation, of which we are a part, has not been formed in parts. The organ of vision, and the matter or influence carried to the organ, and the qualities of bodies with which we are acquainted through it, are parts of a system great beyond our imperfect comprehension, formed as it should seem at once in wisdom; not pieced together like the work of human ingenuity.

When this whole was created, (of which the remote planetary system, as well as our bodies, and the objects more familiar to our observation, are but parts), the mind was placed in a body not merely suited to its residence, but in circumstances to be moved by the materials around it; and the capacities of the mind, and the powers of the organs, which are as a medium betwixt the mind and the external world, have an original constitution framed in relation to the qualities of things.

It is admitted that neither bodies nor the images of bodies enter the brain. It is indeed impossible to believe that colour can be conveyed along a nerve; or the vibration in which we suppose sound to consist can be retained in the brain: but we can conceive, and have reason to believe, that an impression is made upon the organs of the outward senses when we see, or hear, or taste.

In this inquiry it is most essential to observe, that while each organ of sense is provided with capacity of receiving certain changes to be played upon it, as it were, yet each is utterly incapable of receiving the impressions destined for another organ of sensation.

It is also very remarkable that an impression made on two different nerves of sense, though with the same instrument, will produce two distinct sensations; and the ideas resulting will only have relations to the organ affected.

As the announcing of these facts forms a natural introduction to the Anatomy of the Brain, which I am about to deliver, I shall state them more fully.

There are four kinds of Papillae on the tongue, but with two of those only we have to do at present. Of these, the Papillae of one kind form the seat of the sense of taste; the other Papillae (more numerous and smaller) resemble the extremities of the nerves in the common skin, and are the organs of touch in the tongue. When I take a sharp steel point, and touch one of these Papillae, I feel the sharpness. The sense of touch informs me of the shape of the instrument. When I touch a Papilla of taste, I have no sensation similar to the former. I do not know that a point touches the tongue, but I am sensible of a metallic taste, and the sensation passes backward on the tongue.

In the operation of couching the cataract, the pain of piercing the retina with a needle is not so great as that which proceeds from a grain of sand under the eyelid. And although the derangement of the stomach sometimes marks the injury of an organ so delicate, yet the pain is occasioned by piercing the outward coat, not by the affection of the expanded nerve of vision.

If the sensation of light were conveyed to us by the retina, the organ of vision, in consequence of that organ being as much more sensible than the surface of the body as the impression of light is more delicate than that pressure which gives us the sense of touch; what would be the feelings of a man subjected to an operation in which a needle were pushed through the nerve. Life could not bear so great a pain.

But there is an occurrence during this operation on the eye, which will direct us to the truth: when the needle pierces the eye, the patient has the sensation of a spark of fire before the eye.

This fact is corroborated by experiments made on the eye. When the eye-ball is pressed on the side, we perceive various coloured light. Indeed the mere effect of a blow on the head might inform us, that sensation depends on the exercise of the organ affected, not on the impression conveyed to the external organ; for by the vibration caused by the blow, the ears ring, and the eye flashes light, while there is neither light nor sound present.

It may be said, that there is here no proof of the sensation being in the brain more than in the external organ of the sense. But when the nerve of a stump is touched, the pain is as if in the amputated extremity. If it be still said that this is no proper example of a peculiar sense existing without its external organ, I offer the following example: Quando penis glandem exedat ŭlcŭs, et nihil nisi granulatio maneat, ad extremam tamen nervi pudicae partem ubi terminatŭr sensus supersunt, et exquisitissima sensŭs gratificatio.

If light, pressure, galvanism, or electricity produce vision, we must conclude that the idea in the mind is the result of an action excited in the eye or in the brain, not of any thing received, though caused by an impression from without. The operations of the mind are confined not by the limited nature of things created, but by the limited number of our organs of sense. By induction we know that things exist which yet are not brought under the operation of the senses. When we have never known the operation of one of the organs of the five senses, we can never know the ideas pertaining to that sense; and what would the effect on our minds, even constituted as they now are, with a superadded organ of sense, no man can distinctly imagine.

As we are parts of the creation, so God has bound us to the material world by this law of our nature, that it shall require excitement from without, and an operation produced by the action of things external to rouse our faculties: But that once brought into activity, the organs can be put in exercise by the mind, and be made to minister to the memory and imagination, and all the faculties of the soul.

I shall hereafter shew, that the operations of the mind are seated in the great mass of the cerebrum, while the parts of the brain to which the nerves of sense tend, strictly form the seat of the sensation, being the internal organs of sense. These organs are operated upon in two directions. They receive the impression from without, as from the eye and ear: and as their action influences the operations of the brain producing perception, so are they brought into action and suffer changes similar to that which they experience from external pressure by the operation of the will; or, as I am now treating of the subject anatomically, by the operation of the great mass of the brain upon them.

In all regulated actions of the muscles we must acknowledge that they are influenced through the same nerves, by the same operation of the sensorium. Now the operations of the body are as nice and curious, and as perfectly regulated before Reason has sway, as they are at any time after, when the muscular frame might be supposed to be under the guidance of sense and reason. Instinctive motions are the operations of the same organs, the brain and nerves and muscles, which minister to reason and volition in our mature years. When the young of any animal turns to the nipple, directed by the sense of smelling, the same operations are performed, and through the same means, as afterwards when we make an effort to avoid what is noxious, or desire and move towards what is agreeable.

The operations of the brain may be said to be three-fold: 1. The frame of the body is endowed with the characters of life, and the vital parts held together as one system through the operation of the brain and nerves; and the secret operations of the vital organs suffer the controul of the brain, though we are unconscious of the thousand delicate operations which are every instant going on in the body. 2. In the second place, the instinctive motions which precede the developement of the intellectual faculties are performed through the brain and nerves. 3. In the last place, the operation of the senses in rouzing the faculties of the mind, and the exercise of the mind over the moving parts of the body, is through the brain and nerves. The first of these is perfect in nature, and independent of the mind. The second is a prescribed and limited operation of the instrument of thought and agency. The last begins by imperceptible degrees, and has no limit in extent and variety. It is that to which all the rest is subservient, the end being the calling into activity and the sustaining of an intellectual being.

Thus we see that in as far as is necessary to the great system, the operation of the brain, nerves, and muscles are perfect from the beginning; and we are naturally moved to ask, Might not the operations of the mind have been thus perfect and spontaneous from the beginning as well as slowly excited into action by outward impressions? Then man would have been an insulated being, not only cut off from the inanimate world around him, but from his fellows; he would have been an individual, not a part of a whole. That he may have a motive and a spring to action, and suffer pain and pleasure, and become an intelligent being, answerable for his actions,—sensation is made to result from external impression, and reason and passion to come from the experience of good and evil; first as they are in reference to his corporeal frame, and finally as they belong to the intellectual privations and enjoyments.

THE brain is a mass of soft matter, in part of a white colour, and generally striated; in part of a grey or cineritious colour, having no fibrous appearance. It has grand divisions and subdivisions: and as the forms exist before the solid bone incloses the brain; and as the distinctions of parts are equally observable in animals whose brain is surrounded with fluid, they evidently are not accidental, but are a consequence of internal structure; or in other words they have a correspondence with distinctions in the uses of the parts of the brain.

On examining the grand divisions of the brain we are forced to admit that there are four brains. For the brain is divided longitudinally by a deep fissure; and the line of distinction can even be traced where the sides are united in substance. Whatever we observe on one side has a corresponding part on the other; and an exact resemblance and symmetry is preserved in all the lateral divisions of the brain. And so, if we take the proof of anatomy, we must admit that as the nerves are double, and the organs of sense double, so is the brain double; and every sensation conveyed to the brain is conveyed to the two lateral parts; and the operations performed must be done in both lateral portions at the same moment.

I speak of the lateral divisions of the brain being distinct brains combined in function, in order the more strongly to mark the distinction betwixt the anterior and posterior grand divisions. Betwixt the lateral parts there is a strict resemblance in form and substance: each principal part is united by transverse tracts of medullary matter; and there is every provision for their acting with perfect sympathy. On the contrary, the *cerebrum*, the anterior grand division, and the *cerebellum* the posterior grand division, have slight and indirect connection. In form and division of parts, and arrangement of white and grey matter, there is no resemblance. There is here nothing of that symmetry and correspondence of parts which is so remarkable betwixt the right and left portions.

I have found evidence that the vascular system of the cerebellum may be affected independently of the vessels of the cerebrum. I have seen the whole surface of the cerebellum studded with spots of extravasated blood as small as pin heads, so as to be quite red, while no mark of disease was upon the surface of the cerebrum. The action of vessels it is needless to say is under the influence of the parts to which they go; and in this we have a proof of a distinct state of activity in the cerebrum and cerebellum.

From these facts, were there no others, we are entitled to conclude, that in the operations excited in the brain there cannot be such sympathy or corresponding movement in the cerebrum and cerebellum as there is betwixt the lateral portions of the cerebrum; that the anterior and posterior grand divisions of the brain perform distinct offices.

In examining this subject further, we find, when we compare the relative magnitude of the cerebrum to the other parts of the brain in man and in brutes, that in the latter the cerebrum is much smaller, having nothing of the relative magnitude and importance which in man it bears to the other parts of the nervous system; signifying that the cerebrum is the seat of those qualities of mind which distinguish man. We may observe also that the posterior grand division, or *cerebellum* remains more permanent in form: while the cerebrum changes in conformity to the organs of sense, or the endowments of the different classes of animals. In the inferior animals, for example, where there are two external organs of the same sense, there is to be found two distinct corresponding portions of cerebrum, while the cerebellum corresponds with the frame of the body.

In thinking of this subject, it is natural to expect that we should be able to put the matter to proof by experiment. But how is this to be accomplished, since any experiment direct upon the brain itself must be difficult, if not impossible?—I took this view of the subject. The *medulla spinalis* has a central division, and also a distinction into anterior and posterior fasciculi, corresponding with the anterior and posterior portions of the brain. Further we can trace down the crura of the *cerebrum* into the anterior fasciculus of the spinal marrow, and the crura of the *cerebellum* into the posterior fasciculus. I thought that here I might have an opportunity of touching the *cerebellum*, as it were, through the posterior portion of the spinal marrow, and the cerebrum by the anterior portion. To this end I made experiments which, though they were not conclusive, encouraged me in the view I had taken.

I found that injury done to the anterior portion of the spinal marrow, convulsed the animal more certainly than injury done to the posterior portion; but I found it difficult to make the experiment without injuring both portions.

Next considering that the spinal nerves have a double root, and being of opinion that the properties of the nerves are derived from their connections with the parts of the brain, I thought that I had an opportunity of putting my opinion to the test of experiment, and of proving at the same time that nerves of different endowments were in the same cord, and held together by the same sheath.

On laying bare the roots of the spinal nerves, I found that I could cut across the posterior fasciculus of nerves, which took its origin from the posterior portion of the spinal marrow without convulsing the muscles of the back; but that on touching the anterior fasciculus with the point of the knife, the muscles of the back were immediately convulsed.

Such were my reasons for concluding that the cerebrum and the cerebellum were parts distinct in function, and that every nerve possessing a double function obtained that by having a double root. I now saw the meaning of the double connection of the nerves with the spinal marrow; and also the cause of that seeming intricacy in the connections of nerves throughout their course, which were not double at their origins.

The spinal nerves being double, and having their roots in the spinal marrow, of which a portion comes from the cerebrum and a portion from the cerebellum, they convey the attributes of both grand divisions of the brain to every part; and therefore the distribution of such nerves is simple, one nerve supplying its destined part. But the nerves which come directly from the brain, come from parts of the brain which vary in operation; and in order to bestow different qualities on the parts to which the nerves are distributed, two or more nerves must be united in their course or at their final destination. Hence it is that the 1st nerve must have branches of the 5th united with it: hence the *portio dura* of the 7th pervades everywhere the bones of the cranium to unite with the extended branches of the 5th: hence the union of the 3rd and 5th in the orbit: hence the 9th and 5th are both sent to the tongue: hence it is, in short, that no part is sufficiently supplied by one single nerve, unless that nerve be a nerve of the spinal marrow, and have a double root, a connection (however remotely) with both the cerebrum and cerebellum.

Such nerves as are single in their origin from the spinal marrow will be found either to unite in their course with some other nerve, or to be such as are acknowledged to be peculiar in their operation.

The 8th nerve is from the portion of the *medulla oblongata*[1] which belongs to the cerebellum: the 9th nerve comes from the portion which belongs to the cerebrum. The first is a nerve of the class called Vital nerves, controuling secretly the operation of the body; the last is the Motor nerve of the tongue, and is an instrument of volition. Now the connections formed by the 8th nerve in its course to the viscera are endless; it seems nowhere sufficient for the entire purpose of a nerve; for everywhere it is accompanied by others, and the 9th passes to the tongue, which is already profusely supplied by the 5th.

Understanding the origin of the nerves in the brain to be the source of their powers, we look upon the connections formed betwixt distant nerves, and upon the combination of nerves in their passage, with some interest; but without this the whole is an unmeaning tissue. Seeing the seeming irregularity in one subject, we say it is accident; but finding that the connections never vary, we say only that it is strange, until we come to understand the necessity of nerves being combined in order to bestow distinct qualities on the parts to which they are sent.

The *cerebellum* when compared with the *cerebrum* is simple in its form. It has no internal tubercles or masses of cineritious matter in it. The medullary matter comes down from the cineritious cortex, and forms the *crus*; and the *crus* runs into union with the same process from the cerebrum; and they together form the *medulla spinalis*, and are continued down into the spinal marrow; and these crura or processes afford double origin to the double nerves of the spine. The nerves proceeding from the Crus Cerebelli go everywhere (in seeming union with those from the Crus Cerebri); they unite the body together, and controul the actions of the bodily frame; and especially govern the operation of the viscera necessary to the continuance of life.

In all animals having a nervous system, the *cerebellum* is apparent, even though there be no *cerebrum*. The cerebrum is seen in such tribes of animals as have organs of sense, and it is seen to be near the eyes, or principal organ of sense; and sometimes it is quite separate from the *cerebellum*.

The cerebrum I consider as the grand organ by which the mind is united to the body. Into it all the nerves from the external organs of the senses enter; and from it all the nerves which are agents of the will pass out.

---

[1] The *medulla oblongata* is only the commencement of the spinal marrow.

If this be not at once obvious, it proceeds only from the circumstance that the nerves take their origin from the different parts of the brain; and while those nerves are considered as simple cords, this circumstance stands opposed to the conclusion which otherways would be drawn. A nerve having several roots, implies that it propagates its sensation to the brain generally. But when we find that the several roots are distinct in their endowments, and are in respect to office distinct nerves; then the conclusion is unavoidable, that the portions of the brain are distinct organs of different functions.

To arrive at any understanding of the internal parts of the cerebrum, we must keep in view the relation of the nerves, and must class and distinguish the nerves, and follow them into its substance. If all ideas originate in the mind from external impulse, how can we better investigate the structure of the brain than by following the nerves, which are the means of communication betwixt the brain and the outward organs of the senses?

The nerves of sense, the olfactory, the optic, the auditory, and the gustatory nerve, are traced backwards into certain tubercles or convex bodies in the base of the brain. And I may say, that the nerves of sense either form tubercles before entering the brain, or they enter into those convexities in the base of the *cerebrum*. These convexities are the constituent parts of the cerebrum, and are in all animals necessary parts of the organs of sense: for as certainly as we discover an animal to have an external organ of sense, we find also a medullary tubercle; whilst the superiority of animals in intelligence is shewn by the greater magnitude of the hemispheres or upper part of the cerebrum.

The convex bodies which are seated in the lower part of the cerebrum, and into which the nerves of sense enter, have extensive connexion with the hemispheres on their upper part. From the medullary matter of the hemispheres, again, there pass down, converging to the crura, Striae, which is the medullary matter taking upon it the character of a nerve; for from the Crura Cerebri, or its prolongation in the anterior Fasciculi of the spinal marrow, go off the nerves of motion.

But with these nerves of motion which are passing outward there are nerves going inwards; nerves from the surfaces of the body; nerves of touch; and nerves of peculiar sensibility, having their seat in the body or viscera. It is not improbable that the tracts of cineritious matter which we observe in the course of the medullary matter of the brain, are the seat of such peculiar sensibilities; the organs of certain powers which seem resident in the body.

As we proceed further in the investigation of the function of the brain, the discussion becomes more hypothetical. But surely physiologists have been mistaken in supposing it necessary to prove sensibility in those parts of the brain which they are to suppose the seat of the intellectual operations. We are not to expect the same phenomena to result from the cutting or tearing of the brain as from the injury to the nerves. The function of the one is to transmit sensation; the other has a higher operation. The nature of the organs of sense is different; the sensibilities of the parts of the body are very various. If the needle piercing the retina during the operation of couching gives no remarkable pain, except in touching the common coats of the eye, ought we to imagine that the seat of the higher operations of the mind should, when injured, exhibit the same effects with the irritation of a nerve? So far therefore from thinking the parts of the brain which are insensible, to be parts inferior (as every part has its use), I should even from this be led to imagine that they had a higher office. And if there be certain parts of the brain which are insensible, and other parts which being injured shake the animal with convulsions exhibiting phenomena similar to those of a wounded nerve, it seems to follow that the latter parts which are endowed with sensibility like the nerves are similar to them in function and use; while the parts of the brain which possess no such sensibility are different in function and organization from the nerves, and have a distinct and higher operation to perform.

If in examining the apparent structure of the brain, we find a part consisting of white medullary Striae and fasciculated like a nerve, we should conclude that as the use of a nerve is to transmit sensation, not to perform any more peculiar function, such tracts of matter are media of communication, connecting the parts of the brain; rather than the brain itself performing the more peculiar functions. On the other hand, if masses are found in the brain unlike the matter of the nerve, and which yet occupy a place guarded as an organ of importance, we may presume that such parts have a use different from that of merely conveying sensation; we may rather look upon such parts as the seat of the higher powers.

Again, if those parts of the brain which are directly connected with the nerves, and which resemble them in structure, give pain when injured, and occasion convulsion to the animal as the nerves do when they are injured; and if on the contrary such parts as are more remote from the nerves, and of a different structure, produce no such effect when injured, we may conclude, that the office of the latter parts is more allied to the intellectual operations, less to mere sensation.

I have found at different times all the internal parts of the brain diseased without loss of sense; but I have never seen disease general on the surfaces of the hemispheres without derangement or oppression of the mind during the patient's life. In the case of derangement of mind, falling into lethargy and stupidity, I have constantly found the surface of the hemispheres dry and preternaturally firm, the membrane separating from it with unusual facility.

If I be correct in this view of the subject, then the experiments which have been made upon the brain tend to confirm the conclusions which I should be inclined to draw from strict anatomy; viz. that the cineritious and superficial parts of the brain are the seat of the intellectual functions. For it is found that the surface of the brain is totally insensible, but that the deep and medullary part being wounded the animal is convulsed and pained.

At first it is difficult to comprehend, how the part to which every sensation is referred, and by means of which we become acquainted with the various sensations, can itself be insensible; but the consideration of the wide difference of function betwixt a part destined to receive impressions, and a part which is the seat of intellect, reconciles us to the phenomenon. It would be rather strange to find, that there were no distinction exhibited in experiments on parts evidently so different in function as the organs of the senses, the nerves, and the brain. Whether there be a difference in the matter of the nervous system, or a distinction in organization, is of little importance to our enquiries, when it is proved that their essential properties are different, though their union and co-operation be necessary to the completion of their function—the developement of the faculties by impulse from external matter.

All ideas originate in the brain: the operation producing them is the remote effect of an agitation or impression on the extremities of the nerves of sense; directly they are consequences of a change or operation in the proper organ of the sense which constitutes a part of the brain, and over these organs, once brought into action by external impulse, the mind has influence. It is provided, that the extremities of the nerves of the senses shall be susceptible each of certain qualities in matter; and betwixt the impression of the outward sense, as it may be called, and the exercise of the internal organ, there is established a connection by which the ideas excited have a permanent correspondence with the qualities of bodies which surround us.

From the cineritious matter, which is chiefly external, and forming the surface of the cerebrum; and from the grand center of medullary matter of the cerebrum, what are called the *crura* descend. These are fasciculated processes of the cerebrum, from which go off the nerves of motion, the nerves governing the muscular frame. Through the nerves of sense, the *sensorium* receives impressions, but the will is expressed through the medium of the nerves of motion. The secret operations of the bodily frame, and the connections which unite the parts of the body into a system, are through the cerebellum and nerves proceeding from it.

THE END

ABOUT THE AUTHOR

Michael Aminoff is professor of neurology in the School of Medicine at the University of California, San Francisco. He is well known as a neurologist, clinical neurophysiologist, and clinical investigator, and also as an author and medical historian. His books on various neurological topics have been translated into eight languages, and he has also edited several comprehensive textbooks that have gone into numerous editions. His biography of Charles-Édouard Brown-Séquard was published by Oxford University Press in 2011. He lives in San Francisco with his wife, Jan.

INDEX

Page numbers followed by *f* indicate figures. Pages followed by n indicate footnotes, with the specific footnote indicated by a capital letter

Abdominal wounds, 225
Abernethy, John, 20, 20nF, 22, 79, 80*f*, 146
Adam, Frederick, 194, 194nE
Adams, John, 124
Ainslie Place residence, 189
Allen, David, 12, 12nK
Amputation, 69–70
Amyotrophic lateral sclerosis (ALS), 134–135
Anatomy Act, 31, 31nB, 33nC, 153
The Anatomy and Philosophy of Expression as Connected with the Fine Arts (Bell), 73*f*
The Anatomy of Expression (Bell), 20, 24, 29, 33–34, 31nnA–B, 195. *see also Essays on the Anatomy of Expression in Painting* (Bell)
The Anatomy of the Brain, Explained in a Series of Engravings (Bell), 14, 202
The Anatomy of the Human Body (Bell), 13–14, 209–210
Animal expressions, 49
Animal Mechanics (Bell), 146–148, 215
Annals of the Fine Arts, 42
Antithesis, principle of, 47
Antlantoaxial dislocation, 138–139
Apothecaries' Act of 1815, 166
Arris lectures (Arris and Gale lectures), 91, 91nJ
The Artist (Carlisle), 42
Auzoux, Louis, 33nG

Babbage, Charles, 153–154, 157
Baillie, Matthew, 19–20, 19nC, 26, 79
Banks, Joseph, 20, 20nH

Barclay, John, 9, 71–72, 71nnN–O, 72*f*
Barcleian Museum, 71, 71nN–O, 72*f*
Barnes, Thomas, 81
The Barque of Dante (Delacroix), 50
Bausch, Johann Laurentius, 139
Beechey, William, 42, 42nL
Bell, Alexander, 77
Bell, Barbara (Shaw), 77
Bell, Barbara (Wright), 77
Bell, Charles
  award/honors, 2, 117, 157–161, 157nnN–Q, 158–159*f*, 161nR
  background, 6–10
  books by, 209–217
  character, friends and, 15, 82–83, 99–100, 145–146, 157–158, 203
  childhood, 10–13, 11nI
  death of, 195–196, 196nG
  drawings and paintings, list of, 223–226
  education, 13–16, 13nL
  family background, 6–9, 6nD, 9nnE–F, 10nG
  finances, 191, 192*f*, 193, 203
  health issues, 201–202
  Idea of a New Anatomy of the Brain, reprint of, 227–236
  legacy, 201–207, 203nnA–B, 206*f*
  marriage, 26, 77, 79–82
  medical training, writings (early), 13–15, 13nM
  portraits of, 28*f*, 160*f*, 183*f*
  professional work, 83–87, 83nD, 92–93*f*
  published papers and lectures, list of, 219–222

239

Bell, Charles (*Cont.*)
  retirement, return to Edinburgh, 189–197, 190f, 191nD
  Royal Academy of Arts, 20, 25, 28f, 29, 31, 32nC, 41–43, 43nM, 51, 78, 80f, 81
  School of Anatomy, Leicester Square, London, 22–24, 29
  University College London (University of London), 167–180 (*see also* University College London)
Bell, George, 8, 15, 23, 26, 77, 88, 195
Bell, George Joseph, 10
Bell, John (brother), 8–10, 10nG, 13, 15, 87–89
Bell, John (son), 77–78
Bell, Lilias (Grahame), 7
Bell, Margaret (Morice), 7, 11–13, 14
Bell, Marion (Shaw), 26, 77–79, 189–191
Bell, Robert, 8, 15, 87
Bell, William, 7–8, 10
Bell's muscle, 84
Bell's palsy, 132–133
Bell's respiratory nerves, 107, 108, 128, 132, 194
Bell's sign, 132
Bell's spasms, 133
Bentham, Jeremy, 153, 153nI
Bernard, Claude, 101nB
Blicke, Charles, 84
Blizard, William, 19–22, 19nD
Blücher, Gebhard Leberecht von, 65, 70
Bonaparte, Napoleon, 65, 70nM
Brain, 117–121, 117nA, 120nB, 120nC, 118f, 119f. *see also* nervous system function studies
Brewster, David, 160
Bridgewater treatises, 149–154, 150nC, 151nnD–H, 153nI, 194
British Museum, 226
Brougham, Henry, 11, 83, 83nC, 146–148, 156, 167
Brown-Séquard, Charles-Édouard, 120, 120nB, 127–128
Buccinator muscle, 108nI
Buckland, William, 151, 151nF
The Burial of Sir John Moore After Corunna, 58nB

Burke, William, 71nO, 72f
Cadavers, acquisition of, 31, 72f, 153
Campbell, Thomas, 167, 167nA
Cannon-shot wounds, 67f, 225
Carlisle, Anthony, 20, 20nI, 41–43, 43nM, 80f, 81, 146
Cerebellum, 97, 99–100, 100nA, 118f, 119f, 120
Cerebrum, 78, 97–100, 118f, 119f, 120, 121, 204, 227–236
Certificates of attendance, 89f, 173, 175f, 184
Cervical dystonia (torticollis), 137, 138nB
Chalmers, Thomas, 150
Chambers, Robert, 155nL
Charles I, 7
Cheyne, John, 83
Christ in the House of His Parents (Millais), 51
Circle of nerves, 125, 126
Circulatory system studies, 90, 145
Cockburn, Henry Thomas, 83
Conolly, John, 174
Cooper, Astley, 19, 19nB, 22, 42, 79, 80f, 92–94, 146, 183, 191, 195
Corpses, illegal trade in, 31, 69f, 153
Corunna
  Bell's work, 62–65, 223–224
  medical conditions, 59–62, 59nnD–E, 60nF–G
  retreat to, 57–59, 57nA, 58nnB–C
Cranial nerves, 106–112, 107nH, 109–110f
Creation of Adam (Michelangelo), 31nA
Creationism, 4, 44, 44nN, 49, 145–146, 155, 156, 189, 194, 205. *see also* Bridgewater treatises; intelligent design theory
Cromwell, Oliver, 7
Crosse, John, 83
Cruikshank, William, 26, 26nQ
Cuvier, Georges, 33nG, 146–147, 154–155, 156, 168nF

Darwin, Charles, 3, 41, 44–49, 44nN, 113, 145, 154–155, 159f
Darwin, Erasmus, 154, 154nJ
De Humani Corporis Fabrica (Vesalius), 31
Delacroix, Eugène, 50

Dental Hospital of London, 81
Descartes, Rene, 125
The Descent of Man (Darwin), 44
The Disasters of War (Goya), 59
Dissection procedures, 14
Dissertation on Gun-Shot Wounds (Bell), 2, 64–65, 78, 214
Duchenne, Guillaume, 45, 45nO
Duchenne–Aran type spinal muscular atrophy, 45nO
Duchenne muscular dystrophy, 45nO
Dysentery, 61–62
Dystonias, 137–139, 139nB

Eastlake, Elizabeth, 52
Écorchés, 31
Edinburgh School of Arts, 148, 148nB
Edinburgh University, 4, 5–6, 5nA, 8, 9, 9nE–F, 15, 15nN, 71nnN–O, 87, 131, 155, 165, 181–182, 195, 195nF
Egerton, Francis Henry, 150–151
Ekman, Paul, 49
Electrodiagnosis, 45nO
Electrolysis, 20nI
Electrotherapy, 45nO
Essays on the Anatomy of Expression in Painting (Bell)
  on animal forms, 40
  background, 2–3, 33–34, 145
  on beauty of expression, 38–40
  Charles Darwin and, 41, 44–49
  editions of, 34, 34nH, 212
  eyes, 37–38
  facial expressions, 36–40
  facial expressions, artistic representation of, 49–50
  facial expressions, scientific studies of, 44–49, 44nN, 45nO, 46–47f
  illustrations in, 34
  infants, expression of, 38–39
  natural theology theme, 35, 39–40
  on object of painting, 38
  on posture and action, 40, 40nJ
  skull and face changes with age, 35, 36–37f
  smiles, 36
  states of mind, 38, 39f
  success of, 40–41, 165
Evolution (Darwinian), 44–50, 154–156, 154nK, 155nl
Existence of God, 146–147
An Exposition of the Natural System of the Nerves of the Human Body (Bell), 90–91, 102f, 104nF, 214–215
The Expression of the Emotions in Man and Animals (Darwin), 44–48, 145

Facial (seventh) nerve, 107–112, 107nH, 108nI, 109–110f, 112nJ
Facial palsy, 131–133
Field ambulances, hospitals, 70
First aid, 70
Flaxman, John, 41, 49
Fludyer Street, Westminster, 19, 19nA, 193
Fothergill, John, 139, 139nD
Fuseli, Henry, 41, 49

Gale lectures (Arris and Gale lectures), 91, 91nJ
Gall, Franz Joseph, 87, 87nH
Genitourinary surgery, 83, 83nD
George III, 6nB, 7, 19nnA–C, 25nO, 29, 34nI, 59nD, 157, 189nB
George IV, 19nB, 59nE, 83, 117, 147, 157
Géricault, Théodore, 50–51
Glycogen, 101nB
Goldsmid, Isaac, 159f, 167, 167nB
Goya, Francisco, 59
Grant, Robert E., 155, 170, 180
Great Windmill Street (Hunterian) School
  background, 21f, 25–29, 25nO
  Bailie's administration of, 19
  Bell's purchase of, 26, 84, 165
  Bell's teaching at, 26–28, 91–92, 113, 145
  certificate of attendance, 89f
  lecture hall, 21f
  Middlesex Hospital and, 180
  sale, closing of, 29
  students, famous, 70, 78–79, 83, 107
Green, Joseph Henry, 43, 146, 157nO
Gregory, James, 9
Grisel, Pierre, 138
Gunshot wounds, 24nL, 27, 63–66, 67f, 67–71, 74, 123nF, 223–224
Guthrie, George James, 69

Halford, Henry, 189, 189nB
Hall, Marshall, 126
Haslar Hospital, 61–63, 62nnH–I
Hawkins, Caesar, 27, 89f, 107
Haydon, Benjamin, 32, 32nD, 42, 50
Head, Henry, 141, 141nE
Heinrich, Wilhelm Karl (Baron Driesen), 86, 86nG
Hemifacial spasm, 133
Herschel, John, 150, 160, 161nR
Herschel, William, 161nR
Horner, Francis, 11, 11nJ, 66, 66nK, 83
Horner, Leonard, 148, 168, 168nE, 175–178, 175f
Hospital for Women, 81
Hunter, John, 20nG, 22, 25nO, 59nE, 140
Hunter, William, 25–26, 29, 31
Hunterian Museum, 25nO
Hunterian School. *see* Great Windmill Street (Hunterian) School
Hydrophobia, 190f

Idea of a New Anatomy of the Brain (Bell), 1, 90, 99–100, 102, 102nD, 112–114, 121, 122, 213, 227–236
Illustrations of the Great Operations of Surgery (Bell), 90, 214
Imperial Leopoldina Academy of Natural Sciences, 139, 139nC
Institutes of Surgery (Bell), 193–194, 216–217
Intelligent design theory, 146–147, 152, 156, 177, 180, 205. *see also* Bridgewater treatises
Ivory, John, 160

Jefferson, Thomas, 124
Jeffrey, Francis, 33, 33nF, 81, 82, 191nD, 196

Keate, Thomas, 59, 59nE
Keith, Arthur, 203–204, 203nA
Kidd, John, 150, 150nC
King's College London, 27, 107, 111, 146, 171nH, 179
Kirby, William, 151, 151nG

Knight, Francis, 59, 60
Knox, Robert, 68, 71, 72f
Konig, Charles, 160

Laennec, René, 124
Lamarckian inheritance, 48, 154
Laminectomy, 93–94
Landseer, Edwin, 32, 32nC
Landseer, Thomas, 90
Larrey, Dominique Jean, 70, 70nM
Lawrence, Thomas, 42, 43
Lee, Robert, 112nJ
Leonardo Da Vinci, 133
Leslie, John, 160
Letters Concerning the Diseases of the Urethra (Bell), 83, 83nD, 213
Library of Useful Knowledge, 148
Lind, James, 62, 62nI
Linnaean Society, 81
Liston, Robert, 181, 181nM, 197
Locke, John, 139
Locock, Charles, 189, 189nA
Long thoracic nerve, 128, 204
Lyell, Charles, 155, 155nM, 156, 159f
Lynn, William, 19, 19nE, 22, 22nJ, 79

Macartney, James, 22
Magendie, François, 2, 88, 101–106, 101nB, 102f, 108, 111, 112, 124, 202–203
Manual of Anatomy (Shaw), 88, 103
Maton, William, 34, 34nI, 78, 157nO
Mayo, Herbert, 2, 27, 79, 89f, 94, 102f, 106–112, 112nJ, 113, 114, 124, 132, 150, 180, 202, 203
McGrigor, James, 20nG, 60, 60nG, 62
Mechanics' Institutes, 148, 149f
Meckel, Johann Friedrich, 168, 168nF
Medical Act of 1858, 185
Medical education. *see also* specific hospitals and schools by name
  background, 165–167
  Bell's school, Leicester Street, London, 22–24, 29
  certification reforms, 182–183

Hunterian School (*see* Great Windmill Street (Hunterian) School)
King's College London, 27, 107, 111, 146, 171nH, 179
London College of Medicine, 184
Middlesex Hospital Medical School, 20, 180–181
private schools, 24
University College London, 167–180
University of Cambridge, 24nN
University of London, 167–180
University of Oxford, 24nN
Mental illness, depiction of, 38, 39f, 44, 46–47f
Michelangelo, 31, 31nA
Middlesex Hospital
Bell family appointments to, 77, 78, 86f, 88, 89
Bell's appointment to, 84–87, 85f, 180
fee structure, 171, 171nI
founding of, 84–85, 85f, 85nF
medical school, 20, 180–181
University College, association with, 171
Mitchell, Silas Weir, 123, 123nF
Mitscherlich, Eilert, 117nA
Mona Lisa, 133
Monro, Alexander *(secundus)*, 9, 9nF, 13
Monro-Kellie doctrine, 9nF
Moore, John, 57–59, 57nA, 58nB
Motor neuron diseases, 134–135
Movement, 126–127
Movement disorders, 137–138, 138nB
Müller, Johannes, 122–124
Muscle biopsy tool, 45nO
Muscle innervation, 124–125, 125nH
Muscular dystrophies, 135
Myotonia, 136–137

National Hospital for Diseases of the Heart and Paralysis, 81
Natural theology, 83, 146, 148–151, 154–156, 194, 216. see also Bridgewater treatises; Creationism; Intelligent design theory
Nerves, respiratory, of Bell, 107, 108, 128, 132, 194
Nervous system function studies
antlantoaxial dislocation, 138–139

background, 97–98
beginnings of, 78–79, 98–99
Bell's legacy, 201–206
brain, 117–121, 117nA, 118nnB–C, 119f
cerebellum, 97, 99–100, 100nA, 118f, 119f, 120
cerebrum, 78, 97–100, 118f, 119f, 120, 121, 204, 227–236
classification of nerves, 107–108
discoveries, priorities in, 113–114, 201–203
facial nerves, 107–112, 107nH, 108nI, 109–110f, 112nJ
facial palsy, 131–133
Idea of a New Anatomy of the Brain (Bell), 1, 90, 99–100, 102, 102nD, 112–114, 121, 122, 213, 227–236
long thoracic nerve, 128, 204
Magendie-Bell dispute, 2, 88, 101–106, 101nB, 108, 111, 112, 124, 202–203
motor neuron diseases, 134–135
movement, 126–128
movement disorders, 137–138, 138nB
muscle innervation, 124–125, 125nH
muscular dystrophies, 135
myotonia, 136–137
numb chin syndrome, 2, 133
organization of nervous system, 117–128
peripheral nerves, 121
referred pain, 2, 139–142, 194, 204
sensory physiology, 122–124, 123nE, 123nF
Shaw's work on, 88, 101
smell, sense of, 112
spinal cord, 117–121, 117nA, 120nnB–C, 118–119f
spinal nerve roots, 99–105, 100nA, 101nB, 102f, 103nE, 118f (*see also* Magendie-Bell dispute)
tic douloureux, 139–142
vivisection experiments, 100–102, 101nC, 105, 121
The Ninth Bridgewater Treatise: A Fragment (Babbage), 153–154
Numb chin syndrome, 2, 133
Nurses, 166–167

Observations on Italy (Bell), 10nG, 88
On the Origin of Species (Darwin), 156
Opisthotonus, 71, 73f, 224
Owen, Richard, 154, 154nK, 155

Paley, William, 83, 146
Paley's Natural Theology (Paley), 83, 146, 194, 216
Parthenon Marbles, 40, 40nJ
Pattison, Granville Sharp, 168, 170, 173, 174–177, 178, 180
Paul's Letters to His Kinsfolk (Scott), 68, 68nL
Peel, Robert, 197
Peninsular War, 57–59, 57nA, 58nnB–C. see also Corunna
Pepys, Lucas, 59, 59nD
Percy, Pierre-François, 70
Phantom sensation, 123–124, 123nE
Photography, 45nO, 48
Phrenology, 87, 87nH
Pinel, Philippe, 124
Playfair, John, 13
Plinian Club, 44nN, 155
Practical Essays (Bell), 198, 217
Practice of Military Surgery (Guthrie), 70
Pre-Raphaelites, 50–52, 50nP, 52nQ
Presbyterian Church (Church of Scotland), 7
Prochaska, Jiri, 126
Progressive spinal muscular atrophy, 134
Prout, William, 151, 151nH

Quick, Daniel, 132
The Raft of the Medusa (Géricault), 50

Reciprocal inhibition, 127
Referred pain, 2, 139–142, 194, 204
Reform Act of 1832, 147, 156
Respiratory nerves, of Bell, 107, 108, 128, 132, 194
Reynolds, Joshua, 22, 28f, 29, 41, 50
Richardson, John, 83, 192f, 195
Roget, Mark, 151, 151nE
Roux, Philibert Joseph, 86
Royal Academy of Arts, 20, 25, 28f, 29, 31, 32nC, 41–44, 43nM, 51, 78, 80f, 81

Royal College of Surgeons of Edinburgh, 8, 13, 15nN, 26, 32, 64, 71nN, 84
  Bell's paintings at, 71, 223–224
Royal College of Surgeons of England, 24, 25nO
  Bell's appointment to, 84, 91–92, 91nJ, 92f, 170
  Guthrie's presidency, 70
  reform efforts, 183
  training requirements, 166–167
Royal Guelphic Order of Knighthood, 157–161, 157nnP–Q
Royal High School, Edinburgh, 11, 11nI
Royal Society of Edinburgh, 157, 157nN
Royal Society of London, 157, 157nO, 158–159f
Ruskin, John, 51
Russell, James, 15nN

Sabre wounds, 68–69f, 224–225
Sacks, Oliver, 126
Saint-Hilaire, Étienne Geoffroy, 154–155
Scarlet fever, 78–80
Scott, Walter, 11, 67, 68nL
Scott, William, 83
Scottish Episcopal Church, 7
Sensory physiology, 122–124, 123nnE–F
Separation of Land and Water (Michelangelo), 31nA
Separation of Light from Darkness (Michelangelo), 31nA
A Series of Engravings, Explaining the Course of the Nerves (Bell), 14–15
Serviceable-associated habits, 47–48
Setschenov, Ivan, 127
Shaw, Alexander, 27, 89, 101, 111–114
Shaw, Charles, 77, 78
Shaw, David, 77, 77nA
Shaw, John, 26–28, 68, 88–90, 88nI, 107
Sheldon, John, 41, 41nK
Sherrington, Charles Scott, 127, 127nI
Silver urn gift, 189, 189nnA–C
Smell, sense of, 112
Society for the Diffusion of Useful Knowledge, 153
Society of Apothecaries, 167
Soho Square school, 22–24, 22nK, 79

Soult, Jean-de-Dieu, 58nC
Southwood Smith, Thomas, 153
Specific nerve energies doctrine, 122–124
Spinal cord, 117–121, 117nA 120nB, 121nC, 118*f*
Spinal injury, 93–95
Spinal nerve roots, 99–105, 100nA, 102nB, 102*f*, 103nE, 118*f*. see also Magendie–Bell dispute
Spurzheim, Johann Gaspar, 87, 87nH
Stewart, Dugald, 13
Stone, Frank, 51
Surgeons' Square, 9, 9nE
A Surgical Artist at War (Bell), 71–74, 72*f*
A System of Dissections, Explaining the Anatomy of the Human Body, the Manner of Displaying the Parts, and Their Varieties in Disease (Bell), 13, 210–211
A System of Operative Surgery Founded on the Basis of Anatomy (Bell), 22nJ, 24, 24nL, 64, 205, 212

Tetanus, 66, 68*f*, 224, 225
Thomsen, Julius, 137
Thomson, John, 15nN
Tic douloureux, 139–142
Torticollis (cervical dystonia), 138, 138nB
Trigeminal (fifth) nerve, 102*f*, 106–112, 108nI, 109*f*, 110*f*
Trigeminal neuralgia, 139–142
Tumor, 73*f*, 131, 134, 140, 215
Typhus, 61

University College London (University of London)
  Bell's appointment to, 168–170
  Bell's artworks at, 226
  Bell's resignation, 177–178
  chartering of, 179, 179nnK–L
  class tickets, 175, 175nJ
  course syllabus, 170–171
  diploma awarded by, 166–167
  fee structure, 171
  founding of, 27, 27nR, 124, 167–168, 167nnA–B, 168nnC–F, 169*f*, 171nH
  grievances, 173–174
  hospital, 178–179
  inaugural lecture, 172–173
  Middlesex Hospital, association with, 171
  staff responsibilities, interactions, 174–177, 175*f*
University of Edinburgh, 4, 5–6, 5nA, 8, 9, 9nE–F, 10, 15, 15nN, 87, 131, 155, 165, 181–182, 195, 195nF

Vesalius, Andreas, 31
Vestiges of the Natural History of Creation (Chambers), 155–156, 155nL

Wakley, Thomas, 80*f*, 138, 146, 167, 183–184
Walker, Alexander, 106
Wallace, Alfred Russel, 113, 156
Waller, Augustus Desiré, 203, 203nB
Warburton, Henry, 183–184, 184nN
Waterloo, battle of, 27, 57, 65–74, 65nJ, 67–69*f*, 78, 98, 224–225
Watson, Thomas, 171, 171nH, 180
Wax anatomical models, 32–33, 33nG
Wellesley, Arthur, 59
Westminster Hospital, 20, 22, 70, 79, 84nE
Whewell, William, 151, 151nD
Whitbread, Samuel, 85
Whytt, Robert, 126
Wilkie, David, 32, 32nE, 34, 39*f*, 42, 49, 50, 195
William III (William of Orange), 7
William IV, 19nB, 147, 158, 191
Willis, Thomas, 100nA, 107nH, 227
Wilson, James, 20, 20nG, 21*f*, 26, 84
Working class, secular education of, 147, 148–150, 147nA, 149*f*
Wornum, Ralph Nicholson, 51
Writer's cramp, 2, 137–138
Writers to the Signet, 8, 77

York Hospital, 69, 70

Zumbo, Gaetano Giulio, 32

www.ingramcontent.com/pod-product-compliance
Ingram Content Group UK Ltd.
Pitfield, Milton Keynes, MK11 3LW, UK
UKHW022153230426
12049UKWH00003BA/71